British Wool Marketing Board
Oak Mills, Station Road, Clayton, Bradford, West Yorkshire BD14 6JD
Tel: (0274) 882091, Telex: 51406 BWMB G.

Offering marketing and promotional services to the textile industry for British Wool.

TRADE NAME: British Wool Symbol.

BRANCHES: JAPANESE OFFICE: British Wool Marketing Board, Japan, 605 Royal Wakaba, 21 Wakaba I — Chome, Shinjeku-ku, Tokyo, Japan.

NEW HORIZONS

NEW HORIZONS
With a view to maintaining our market position as a leading supplier of 100% wool and 80% wool/20% nylon carpet yarns, Lawtons have established a long term corporate plan firmly fixed with new horizons. These cover three major areas:

LOCATION
Our new home at Meltham covers 200,000 sq.ft. of high quality modern factory space rationalizing our former three separate factory units under one roof. The streamlining of our production plant increases efficiency and gives potential growth on the site to meet the ever increasing demand for our products in the carpet trade.

PLANT
Systematic planned investment in the latest blending, carding, spinning, twisting and winding machinery is being carried out to create one of the most modern woollen carpet yarn production units in the country.

The use of advance computer technology gives sophisticated control of all aspects in the Company from buying, through the production cycle to despatch of the product.

MARKETING
Our commitment to high quality bespoke yarn production is foremost in our future plans and emphasis will be placed on yarn engineering, research and development of new products which, coupled with our continued attention to quality and service, will enable us to respond even more quickly to the ever changing trends in the market place.

fred lawton & son

Fred Lawton & Son Ltd.,
Meltham Mills, Meltham, Huddersfield,
West Yorks HD7 3AY.
Tel: 0484 852573 Fax: 0484 852737 Telex: 51548

wool tweed
wool THE HEART OF A GOOD CARPET

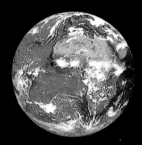

The Kendale Glossary of Basic Textile Terms compiled in five languages and edited in the countries of origin.

Das Kendale Glossary of Basic Textile Terms, verfßt in fünf Sprachen und überarbeitet in den Ursprungsländern.

El Glosario Kendale de Términos Textiles Básicos compilado en cinco idiomas y redactado en los païses de origen.

Le "Glossaire Kendale des Termes textiles Fondamentaux" compilé en cinq langues et édité dans les pays d'origine.

Il "Kendale Glossary of Basic Textile Terms" è stato compilato in cinque lingue e revisionato nei paesi di origine.

GLOSSARY OF BASIC TEXTILE TERMS

GRUNDLEGENDES TEXTILFACHWORTEBUCH

GLOSARIO DE TERMENES BASICOS TEXTILES

GLOSSAIRE DE TERMES TEXTILE DE BASE

GLOSSARIO DI TERMINI BASICI TESSILI

Compiled by J.R. Kenyon, England.
Typeset by Falcon Typographic Art Ltd, Edinburgh, Scotland.
Printed by Gorenjski tisk, Jugoslavia.
Published and distributed by Kendale Publications Ltd.
Huddersfield HD8 9LA, England

Verfasser: J.R. Kenyon, England.
Schriftsatz: Falcon Typographic Art Ltd, Edinburgh, Schottland.
Druck: Gorenjski tisk, Jugoslawien.
Verlag und Vertrieb: Kendale Publications Ltd.
Huddersfield HD8 9LA, England

Compilado por J.R. Kenyon, Inglaterra.
Compuesto por Falcon Typographic Art Ltd, Edinburgh, Escocia.
Impreso por Gorenjski tisk, Yugoslavia.
Publicado y distribuido por Kendale Publications Ltd.
Huddersfield HD8 9LA, Inglaterra

Rédigé par J.R. Kenyon, Angleterre.
Composé par Falcon Typographic Art Ltd, Edinburgh, Ecosse.
Imprimé par Gorenjski tisk, Yougoslavie.
Publié et distribué par Kendale Publications Ltd.
Huddersfield HD8 9LA, Angleterre

Compilazione: J.R. Kenyon, Inghilterra.
Composizione tipografica: Falcon Typographic Art Ltd,
Edinburgh, Scozia
Stampato da: Gorenjski tisk, Yugoslavia.
Casa editrice e distributrice: Kendale Publications Ltd.
Huddersfield HD8 9LA, Inghilterra.

ISBN 09515375 12

ACKNOWLEDGEMENT

The publishers wish to thank all the many friends both in the United Kingdom and abroad for their considerable assistance and patience – over the last three years – in the preparation of this glossary.

DANKSAGUNG

Der verlag möchte auf diesem wege all den vielen freunden in Grossbritannien und im ausland danken, dei bei der erstellung dieses glossars – über die leizien drei jahre – so ausserordentlich hilfreich und geduldig waren.

AGRADECIMIENTOS

Los editores desean dar las gracias a los muchos amigos tanto en el Reino Unido como en últramar por su ayuda y paciencia considerable.

REMERCIEMENTS

Les éditeurs tiennent a remercier tous les nombreux amis, tant du Royaume-Uni que de l'étranger, qui n'ont ménage ni leurs efforts ni leur patience — au cours des trois dernières années — pour élaborer ce.

RICONOSCIMENTO

Gli editori desiderano ringraziare i numerosi amici in Gran Bretagna e all'estero che, negli ultimi tre anni, hanno collaborato alla stesura del presente glossario con encomiabile pazienza.

PREFACE

The concept of The Kendale Glossary of Basic Textile Terms was born out of the onset of the Single European Market and the requirement by the Textile Trade for an international reference work to speed the passage of technical terms and phrases across the barriers of five languages.

Three years of compilation and research have confirmed that the translation of technical words and phrases from one European language to another is a minefield of misunderstanding even between those with a good technical knowledge of the languages concerned. For this very reason the editors have employed expert translators in each country of origin with each translator having an active involvement in the textile industry.

The Glossary has been so arranged that the English language references appear in alphabetical order with four individual indexes allowing other language users to translate with equal facility.

The potential saving in time and money by the correct interpretation of technical instructions and data will be obvious to all manufactures and suppliers engaged in international trade.

It is envisaged that the Kendale Glossary will remain an invaluable work of reference for years to come.

J. Richard Kenyon
Chairman and Managing
Director
Kendale Publications. Ltd.

VORWORT

Das Konzept für das Kendale Glossary of Basic Textile Terms entstand aus den Anfangen des europäischen ßinnenmarktes und dem Bedürfnis des Textilhandels nach einem internationalen Nachschlagwerk, das eine schnellere Übermittlung von Fachausdrücken und Formulierungen über die durch die fünf Sprachen gesetzten Grenzen hinweg ermöglicht.

Die drei Jahre der Begriffssammlung und Forschung haben bestätigt, daß die Übersetzung von Fachausdrücken und Formulierungen aus einer europäischen Sprache in die andere ein regelrechtes Minenfeld der Mißverständnisse sein kann, selbst wenn die Sprecher über gute allgemeine Kenntnisse der betreffenden Sprachen verfügen. Aus eben diesem Grund haben die Herausgeber in jedem Ursprungsland erfahrene Fachübersetzer herangezogen, die aktiv mit der Textilindustrie zu tun haben.

Das Glossar ist so aufgebaut, daß die englischen Schlagwörter in alphabetischer Reihenfolge erscheinen. Daneben gibt es vier Einzelverzeichnisse, damit Benutzer mit anderern Ausgangssprachen ebenso leicht übersetzen können.

Daß man mit der richtigen Interpretation von technischen Anweisungen und Daten Zeit und Geld sparen kann, liegt für alle Hersteller und Lieferanten, die international Hnadel treiben, klar auf der Hand.

Es wird erwartet, daß das Kendale Glossar noch über Jahre hinaus ein wertvolles und unentbehrliches Nachschlagwerk bleiben wird.

J. Richard Kenyon
Vorstand und Geschäftsführer
Kendale Publications Ltd.

PREFACIO

Nació el concepto del Glosario Kendale de Términos Textiles Básicos del comienzo del Mercado Europeo Único y de la necesided para el Comercio textil de una obra de referencia International para edelantor el paso de términos y frases técnicas a través de las barreras de cinco idiomas.

Tres años de complicación e investigación han confirmado que la traducción de palabras y frases técnicas de un idioma europeo a otro es campo de minas de equivocaciones aun entre personas con buen entendimiento pero no-tećnico de los idiomas envueltos. Por esta misma razón, los redactores han empleado a traductores peritos en cada de origen, cada traductor teniendo participación activa en la industria textil.

La disposición del Glosario es tal que las referencias en inglés aparecen en orden alfabético con cuatro indices individuales permitiendo que los usuarios de otros idiomas traduzcan con igual facilidad.

El ahorro potencial de tiempo y dinero debido a la interpretación correcta de instrucciones y datos técnicos será obvio a todos los fabricantes y suministradores que participan en el comercio International.

Se prevé que el Glosario Kendale seguirá siendo una obra de referencia inestimable por muchos años.

J. Richard Kenyon
Presidente y Gerente Director
Kendale Publications Ltd.

PREFACE

Le concept du 'Glossaire Kendale des Termes Textiles' de base est dû à la survenance du Marché Européen Unique et à la demande par l'Industrie Textile d'un ouvrage international de référence visant à accélérer le passage des termes et expressions techniques au-delà de cinq frontières linguistiques.

Trois années de compilation et de recherche ont confirmé que la traduction de mots et d'expressions techniques, d'une langue européenne à une autre, est sujette à toutes sortes d'erreurs même entre ceux qui possèdent une bonne connaissance non technique des langues concernées. C'est pour cette raison que les rédacteurs ont employé des traducteurs-terminologues dans chaque pays d'origine, chaque traducteur jouant un rôle actif dans l'industrie textile.

Le Glossaire a été ainsi élaboré de sorte que les références en langue anglaise soient classées par ordre alphabétique avec quatre index individuels permettant aux autres linguistes de traduire avec autant de facilité.

L'économie potentielle de temps et d'argent, réalisée en interprétant correctement les données et instructions techniques, sera de toute évidence pour les fabricants et les fournisseurs ayant des engagements commerciaux internationaux.

Il est entendu que le 'Glossaire Kendale' restera un ouvrage de reference inestimable pendant les années à venir.

J. Richard Kenyon
Présidente-Directeur Général
Kendale Publications Ltd.

PREFAZIONE

L'idea di compilare il 'Kendale Glossario di Termini Basici Tessili' à stata ispirata dall'avvento del mercato unico europeo e dalla richiesta da parte del settore tessile di un libro di consultazione volto a garantire un rapido scambio di termini e frasti tecniche che abbracci cinque lingue.

Sono occorsi tre anni di compilazione e ricerca, che hanno ampiamente dimostrato come la traduzione di parole e frasi tecniche da una lingua europea ad un'altra sia fonte inesauribile di malintesi, anche tra coloro che hanno una buona padronanza della lingua in questione, ma non in campo tecnico. E'questa la ragione che ha spronato gli editori a richiedere la collaborazione di esperti traduttori residenti nel paese di origine, personalmente coinvolti nel settore tessile.

Il glossario è stato compilato in modo che i termini in lingua inglese compaiano in ordine alfabetico con quattro indici individuali, volti a garantire la medesima facilità di traduzione dalle altre lingue.

Il potenziale risparmio di tempo e denaro garantito dalla corretta interpretazione di istruzioni e dati tecnici è un elemento ben noto a tutti i fabbricatori e i fornitori che operano nel settore internazionale.

Si stima che il glossario Kendale rimarrà un prezioso strumento di consultazione per molti anni a venire.

J. Richard Kenyon
Presidente e Direttore Generale
Kendale Publications Ltd.

Method of Reference

To find the required phrase in the alphabetical Glossary or in the index first look for the collective word, then the descriptive word or words e.g. WEAVING YARN will be entered as YARN-WEAVING, CASHMERE CLOTH will be entered as CLOTH-CASHMERE.

Nachschlagverfahren

Um den gewünchten Begriff im alphabetischen Verzeichnis oder im Index zu finden, schlägt man zuerst den Sam melbegriff nach, dann die nähere(n) Bezeichnung(en). So wird z.B. WEAVING YARN (Webgarn) in der Form YARN-WEAVING und CASHMERE CLOTH (Kaschmirstoff) als CLOTH-CASHMERE eingetragen.

El Método de Consultar

Para encontrar la frase deseada en el Glosario alfabético o en el indice, ante todo buscar la palabra colectiva, luego la palabra o palabras descriptiva(s). p.ej. la frase HILO PARA TEJER será inscrito bajo la palabra colectiva de HILOS. El 'CASIMERE' será inscrito bajo el grupo genérico de TELAS.

Méthode de référence

Pour trouver l'expression requise dans le glossaire alphabétique ou dans l'index, il convient d'abord de chercher le mot-clé et ensuite le ou les sous-mots-clés. Ex: WEAVING YARN sera classé sous YARN-WEAVING, CASHMERE CLOTH sera classé sous CLOTH-CASHMERE.

Metodo di rimando

Per trovare la frase richiesta nel glossario in ordine alfabetico o nell'indice, cercare innanzitutto il nome collettivo, poi il nome o i nomi descrittivi. Ecompio: FILATO PER TESSITURA sarà elencato sotto FILATO-TESSITURA TESSUTO DI CASHMERE sarà elencato sotto CASHMERE-TESSUTO. Il "Kendale Glossary of Basic Textile Terms" è stato compilato in cingue lingue e revisionato nei paesi di origine.

A

1 abaca fibre
Abakafaser (f)
fibra (f) de abaca
fibre (f) d'abaca
fibra (f) di abaca

2 abrade (v)
abscheurern, abreiben
raer, raspar, desgastar por abrasión
ronger par frottement, abraser
abradare

3 abraded yarn
abgeriebenes Garn (n) vom Hersteller
hilado (m) desgastado testurado
en origen
fil (m) abrasé texturé à la filature (f)
filato abraso

4 abrasion
Abrieb (m)
abrasión (f)
abrasion (f)
abrasione (f)

5 abrasion resistance
Scheuerfestigkeit (f)
resistencia (f) a la abrasión
résistance (f) à l'abrasion (f)
resistenza (f) all'abrasione

6 abrasion test
Abriebtest
prueba de abrasión
contrôle (m) d'abrasion
prova (f) di abrasione

7 abrasion tester
Scheuerprüfgerät (n)
ensayador (m) de abrasión,
abrasímetro (m)
brasimètre (m)
abrasiometro (m)

8 absence, lack
Abwesenheit (f), Mangel (m)
Fehlen (n)
ausencia (f), falta (f)
absence (f), manque (m)
assenza (f), mancanza (f)

9 abrasives
Schleifmittel (n pl)
abrasivos (m pl)
abrasifs (m pl)
abrasivi (m pl)

10 absenteeism
Absentismus (m), Abwesenheit (f)
absentismo (m)
absentéisme (m)
assenteismo (m)

11 absorb (v)
aufnehmen, absorbieren
absorber
absorber
assorbire

12 absorbent
Absorptionsmittel (n), Absorber (m)
absorbente (m)
absorbant (m), absorbeur (m)
assorbente (m)

13 absorbent cotton, surgical cotton
medizinische Watte (f)
algodón (m) hidrófilo
coton (m) hydrophile, ouate (f)
hydrophile
cotone idrofilo (m),
ovatta (f)

14 absorptive, absorbing
saugfähig
absorbente
absorbant
assorbente

15 absorption
Absorbtion (f)
absorción
absorption (f)
assorbimento (m)

16 absorption test
Absorbtionstest (m)
ensayo (m) de absorción
contrôle (m) de l'absorption
prova (f) di assorbimento

17 accelerate (v)
beschleunigen
acelerar
accélérer
accelerare (f)

1 acceleration
Beschleunigung (f)
aceleración (f)
accélération (f)
acceleratore

2 accelerator
Beschleuniger (m)
acelerador (m), acelerante (m)
accélérateur (m)
accelerante (m)

3 accessories
Zubehörteile (n pl)
accesorios (m pl)
accessoires (m pl)
accessori (m pl)

4 accountant
Buchhalter (m)
contable (m or f)
comptable (m)
ragioniere (m)

5 accounts
Konten (n pl)
cuentas (f pl)
comptabilite (f)
conti (m pl)

6 accumulation
Anhäufung (f)
aglomeración (f)
accumulation (f)
accumulo (m)

7 acetate
Azetat (n)
acetato (m)
acétate (m)
acetato (m)

8 acetate dye
Azetatfarbstoff (m)
colorante (m) para estrón,
colorante (m) para acetato
colorant (m) pour acétate
colorante (m) per acetato

9 acetate fabric
Azetatgewebe (n)
tejido (m) de acetato
tussu (m) d'acétate
tessuto (m) di acetato

10 acetate fibre
Azetatfaser (f)
fibra (f) de acetato
fibre (f) d'acétate
fibra (f) di acetato

11 acetate rayon
Azetatreyon (n), Azetatseide (f)
rayón (m) al acetato
rayonne (f) acétate
raion (m) acetato

12 acetate staple
Azetatzellwolle (f)
fibra (f) cortada de acetato
fibranne (f) d'acétate
fiocco (m) di acetato

13 acetate yarn
Azetatgarn (n)
hilado (m) de acetato
filé (m) d'acétate
filato (m) di acetato

14 acetic acid
Essigsäure (f)
ácido (m) acético
acide (m) acétique
acido (m) acetico

15 acetone
Azeton (n)
acetona (f)
acétone (m)
acetone (m)

16 acid dye
Säurefarbstoff (m)
colorante (m) ácido
colorant (m) acide
tintura (f) acida

17 acid fastness
Säureechtheit (f)
solidez (f) a los ácidos
solidité (f) aux acides
solidità (f) agli acidi

18 acid fat
Sauerfett (n)
grasa (f) ácida
graisse (f) acide
grasso (m) acido

1 acidify
ansäuern
acidificar
acidifier
acidificare

2 acidity
Azidität (f)
acidez (f)
acidité (f)
acidità (f)

3 acids
Säuren (f pl)
ácidos (m pl)
acides (m pl)
acidi (m pl)

4 acidulated bath, acid liquor
Säurebad (n), Säureflotte (f)
baño (m) ácido
bain (m) acide
bagno (m) acido

5 acid resistance
Säurebeständigkeit (f), Säurefestigkeit (f)
resistencia (f) a los ácidos
résistance (f) aux acides
resistenza (f) agli acidi

6 acid solution
Säurelösung (f)
solución (f) de ácido
solution (f) acide
soluzione (f) acida

7 acrylic
Acryl (n)
acrílico (m)
acrylique (m)
acrilico (m)

8 acrylic acid
Acrylsäure (f)
ácido (m) acrílico
acide (m) acrylique
acido (m) acrilico

9 acrylic fibre
Acrylfaser
fibra (f) acrílica
fibre (f) acrylique
fibra (f) acrilica

10 acrylic fibre and blends
Acrylfasern (f pl) und Stoffe (m pl)
fibra (f) acrílica y mezclas (f pl)
fibre (f) acrylique et mélanges (m pl)
fibre (f pl) e misti (m pl) acrilici

11 acrylic size
Acrylschlichte (f)
cola (f) acrílica
colle (f) acrylique
colla (f) acrilica

12 activate (v)
aktivieren
activar
activer
attivare

13 activator
Aktivierungsmittel (n)
activador
activateur (m)
attivatore (m)

14 activity
Aktivität (f)
actividad (f)
activité (f)
attività (f)

15 actual, real, effective
wirklich, effektiv
efectivo
effectif, réel
effettivo, attuale

16 actuate, run, operate (v)
antreiben, einwirken, in Bewegung setzen
funcionar, actuar
actionner
mettere in movimento

17 adapt, fit, adjust (v)
anpassen
adaptar, ajustar
adapter
adattare, regolare

18 adaptor
Paßstück (n), Adaptor (m)
soporte (m)
adapteur (m)
presa (f) multipla, riduttore (m)

1 add (v)
zusetzen, nachsetzen
añadir
ajouter
aggiungere

2 addition
Zusatz (m), Zusetzen (n)
adición (f)
addition (f), adjonction (f), ajout (m)
aggiunta (f)

3 additive
Zusatz (m)
aditivo (m)
additif (m)
additivo (m)

4 address
Adresse (f)
señas (f pl), dirección (f)
adresse (f)
indirizzo (m)

5 adhere, cling (v)
anhaften, haften
adherir
adhérer
aderire

6 adhesive bonding
adhäsive Vliesverfestigung (f),
adhäsives Bondieren (n)
encolado (m) adhesivo
liaison (f) par adhésif (m)
aderenza (f) adesiva

7 adhesives
Klebstoffe (m pl)
adhesivos (m pl)
adhésifs (m pl)
adesivi (m pl)

8 adjacent fabric
Nachbarstoff (m)
tejido (m) para comparar
tissu (m) témoin
tessuto (m) adiacente

9 adjust, adapt, fit
anpassen
ajustar, adaptar
adapter
adattare, aggiustare

10 adulterate (v)
fälschen
adulterar
adultérer
adulterare

11 advantage
Vorzug (m), Vorteil (m)
ventaja (f)
avantage (m)
vantaggio (m)

12 advantageous
vorteilhaft, nützlich
ventajoso
avantageux
vantaggioso (m)

13 advertising, publicity
Reklame (f), Werbung (f)
publicidad (f), propaganda (f)
publicité (f), réclame (f)
pubblicità (f)

14 aerosols
Aerosole (n pl)
pulverizadores (m pl)
aérosols (m pl)
aerosol (m pl)

15 aerosols – antistatic
Aerosole – antistatisch
pulverizadores (m pl) antiestáticos
aérosols (m pl) – antistatiques
aerosol (m pl) antistatici

16 affect, act (v)
einwirken
obrar
agir
agire

17 affinity
Affinität (f)
afinidad (f)
affinité (f)
affinità (f)

18 aftertreatment
Nachbehandlung (f)
tratamiento (m) posterior
finissage (m)
post-trattamento (m)

1 agents
Wirkstoffe (m pl)
agentes (m pl)
agents (m pl)
agenti (m pl)

2 agitate (v)
schütteln
agitar
agiter
agitare

3 agitation, mixing, stirring
Schütteln (n), Umrühen (n)
agitación (f)
agitation (f), remuage (m)
agitazione (f)

4 agitator, mixer, stirrer
Rührapparat (m), Rührwerk (n)
agitador (m)
agitateur (m)
agitatore (m)

5 agree, correspond (v)
übereinstimmen
corresponder
correspondre
corrispondere

6 air
Luft (f)
aire (m)
air (m)
aria (f)

7 air circulation
Luftzirkulation (f)
circulación (f) de aire
circulation (f) d'air
circolazione (f) d'aria

8 air cleaning
luftreinigend (adj), Luftreinigung (f)
limpieza (f) por aire
épuration (f) d'air
depurazione (f) ad aria

9 air cleaning equipment
Luftreinigunsgeräte (n pl)
equipos (m pl) para purificar el aire
équipement (m) d'épuration d'air
apparecchiature (f pl) di depurazione ad aria

10 air cooling
Luftkühlung (f)
refrigeración (f) por aire, enfriamiento (m) por aire
refroidissement (m) à l'air
raffreddamento (m) ad aria

11 air conditioning
Klimatisierung (f)
acondicionamiento (m) de aire
climatisation (f)
condizionamento (m) d'aria

12 air current
Luftstrom (m)
corriente (f) de aire
courant (m) d'air
corrente (f) d'aria

13 air dried
lufttrocken
secado al aire
sec à l'air
secco ad aria

14 air flow
Luftströmung (f)
caudal (m) de aire
écoulement (m) d'air
flusso (m) d'aria

15 air filter
Luftfiler (m)
filtro (m) de aire
filtre (m) à air
filtro (m) d'aria

16 air heating
Luftheizung (f)
calentamiento (m) de aire
réchauffage (m) d'air
riscaldamento (m) dell'aria

17 air jet
Luftdüsenstuhl (m)
telar (m) con toberas de aire
métier (m) à tisser à tuyères (f pl) (à air (m))
telaio (m) a getto d'aria

18 air splicing
Luftspleißung (f)
refuerzo (m) de aire
renfort (m) pneumatique
rinforzo ad aria

1 airtight
luftdicht (adj)
hermético
à étanche d'air
a chiusa ermetica

2 alcohol
Alkohol (m)
alcohol (m)
alcool (m)
alcool (m)

3 alginate fibre
Alginatfaser (f)
fibra (f) de alginato
fibre (f) d'alginate
fibra (f) d'alginato

4 alignment
Fluchtlinie (f), Ausrichten (n)
alineación (f)
alignement (m)
allineamento (m)

5 alizarin
Alizarin (n)
alizarina (f)
alizarine (f)
alizarina (f)

6 alkali
Alkali (n)
áacali (m)
alcali (m)
alcali (m)

7 alkali fastness
Alkaliechtheir (f)
solidez (f) a los álcalis
solidité (f) aux alcalis
solidità (f) agli alcali

8 alkaline
alkalisch (adj)
alcalino
alcalin
alcalino

9 alkaline solution
Alkalilösung (f)
solución (f) alcalina
solution (f) alcaline
soluzione (f) alcalina

10 alkali resistant
alkalibeständig (adj)
resistente a los álcalis
résistant aux alcalis
resistente agli alcali

11 allow (v)
erlauben
permitir
permettre
permettere

12 allowable
zulässig (adj)
admisible
admissible
ammissibile

13 allowance, tolerance
Toleranz (f)
tolerancia (f)
tolérance (f)
tolleranza (f)

14 alloy
Legierung (f)
aleación
alliage (m)
lega (f)

15 all-wool, pure wool
Reinwolle (f), rein-wollen (adj)
lana (f) pura
pure laine (f)
pura lana (f)

16 alteration
Änderung (f), Alteration (f)
alteración (f)
altération (f)
alterazione (f)

17 alternate
alternativ, abwechselnd (adj)
alternante
alternatif
alternare

18 amber
Bernstein (m)
ambarino (adj)
ambré (adj)
ambra (f)

1 amethyst
Amethyst (m)
de color de amatista
de couleur d'améthyste
ametista

2 American cloth
Wachstuch (n)
hule (m)
toile (f) cirée
tela (f) americana

3 amino acid
Aminosäure (f)
aminoácido (m)
aminoacide (m)
amminoacido (m)

4 ammonia
Ammoniak (m)
amoniaco (m)
ammoniaque (f)
ammoniaca (f)

5 ammonium
Ammonium (n)
amonio (m)
ammonium (m)
ammonio (m)

6 amount, sum
Betrag (m)
importe (m)
montant (m)
importo (m), somma (f)

7 amylopectin
Amylopektin (n)
amilopectina (f)
amylopectine (f)
amilopectina (f)

8 analysis
Analyse (f)
análisis (m)
analyse
analisi (f)

9 analysts
Analytiker (m pl)
analistas (m pl)
analystes (m pl)
analisti (m pl)

10 analytical
analytisch (adj)
analítico
analytique
analitico

11 analyse (v)
analysieren
analizar
analyser
analizzare

12 angle
Winkel (m)
ángulo (m)
angle (m)
angolo (m)

13 angola
Angola (n)
angola
angora
capra d'angora

14 angora
Angora (n)
lana (f) de angora
angora (m)
angora (f)

15 aniline
Anilin (n)
anilina (f)
aniline (f)
anilina (f)

16 animal
Tier (n)
animal (m)
animal (m)
animale (m)

17 animal fat
tierisches Fett (n)
grasa (f) animal
graisse (f) animale
grasso (m) animale

18 animal fibres
tierische Fasern (f pl)
fibras (f pl) animales
fibres (f pl) animales
fibre (f pl) animali

1 animal hair
Tierhaar (n)
pelo (m) animal
poil (m) animal
pelo (m) animale

2 animal wax
tierisches Wachs (n)
cera (f) animal
cire (f) animale
cera (f) animale

3 anticlockwise, counterclockwise
entgegen dem Uhrzeigersinn
en dirección contraria a la de las agujas del reloj
en sens inverse des aiguilles d'une montre
in senso antiorario

4 anticrease
knitterfest (adj)
resistente al arrugado
apprêt (m) infroissable
antipiega

5 anti-felting
Antifilzausrüstung (f)
acabado (m) antienfieltrante
finissage (m) infeutrable, apprêt (m)
infetribile

6 anti-freeze
Gefrierschutz (m)
anticongelante (m)
antigel (m)
antigelo (m)

7 anti-shrink treatment
Antischrumpfbehandlung (f)
tratamiento (m) antiencogible
traitement (m) irrétrécissable
trattamento (m) antiretrazione

8 anti-slip
rutschfest
antideslizante
antiglissant
antisdrucciolevole

9 anti-slip agents
Antirutschmittel (n pl)
agentes (m pl) antideslizantes
agents (m pl) antiglissants
agenti (m pl) antisdrucciolevoli

10 anti-static
antistatisch
antiestático
antistatique
antistatico

11
anti-static agents
Antistatikmittel (n pl)
agentes (m pl) antiestáticos
agents (m pl) antistatiques
agenti (m pl) antistatici

12 apparatus
Apparat (m)
aparato (m)
appareil (m)
apparecchi (m)

13 apparel
Kleidung (f) (adj)
indumentaria (f)
vêtement (m)
abbigliamento

14 appearance
Aussehen (n), Bild (n)
aspecto (m), apariencia (f)
aspect (m)
aspetto (m)

15 apple-green
apfelgrün
verde manzana
vert pomme
verde mela

16 appliance
Vorrichtung (f)
dispositivo (m)
dispositif (m)
strumento, apparecchio (m)

17 application
Anwendung (f)
aplicación (f)
application (f)
applicazione (f)

18 apply to (v)
anwenden
aplicar a
appliquer sur
applicare a

1 appreciable, considerable, significant
erheblich (adj)
notable, considerable, significante
considérable, important
apprezzabile notevole,
significatiuo

2 apprentice
Lehrling (m)
aprendiz (m), aprendiza (f)
apprenti (m)
apprendista (m)

3 appropriate, suitable
passend, sachgemäp
apropiado
convenable
appropriato, adatto

4 approximate
annähernd (adj)
aproximado
approximatif
approsiusiuo – tivo

5 apricot
aprikosenfarben (adj)
albaricoque (m)
abricot (adj inv)
albicocca (f)

6 aprons
Riemchen (n pl)
manguitos (m pl) de estiraje
manchons (m pl)
nastri (m pl) trasportatori a piastre

7 apron cloths
Riemchenstoffe (m pl)
tela (f) para manguitos
étoffe (f) pour tablier
tessuti (m pl) per nastro trasportatore
a piastre

8 apron combs
Riemchenkämme (m pl)
peines (m pl) para manguitos
peignes (f pl) du tablier
pettini (m pl) per nastro trasportatore
a piastre

9 apron leathers
Riemchenleder (n pl)
cueros (m pl) para manguitos
lanières (f pl)
pelli (f pl) per nastro trasportatore
a piastre

10 aprons, spinning
Riemchenspinnen (n)
manguitos (m pl) hilatura
tabliers (m pl) de filature
filatura (f) a nastri trasportatori
a piastre

11 aprons, synthetic
Riemchen (n pl) synthetisch
manguitos (m pl) sintéticos
tabliers (m pl) synthétiques
prodotti (m pl) sintetici per nastri
trasportatori a piastre

12 aquamarine
aquamarinblau
de color de aguamarina
aigue-marine
acquamarina

13 area, zone, surface
Fläche (f), Oberfläche (f)
área (f)
aire (f), région (f), surface (f)
area (f) zona (f),
superficie (f)

14 area weight
Flächengewicht (n)
peso (m) de área
poids (m) superficiel
peso d'aria

15 arm
Arm (m)
brazo (f)
bras (m)
braccio (m)

16 arrange, place, lay (v)
gliedern, anordnen, zusammenstellen,
arrangieren
disponer, colocar
disposer, placer
disporre, collocare

17 arrangement, layout, array, disposition
Anordnung (f), Zusammenstellung (f)
disposición (f)
arrangement (m), disposition (f)
disposizione (f)

18 arras tapestry
Arras-Wandteppich (m)
tapiz (m)
tapisserie (f) d'Arras
arazzo (m)

1 arsenic
Arsen (n)
arsénico (m)
arsenic (m)
arsenico (m)

2 article, item
Artikel (m), Produkt (n), Warenstück (n)
artículo (m)
article (m)
articolo (m)

3 artificial
künstlich (adj)
artificial
artificiel
artificiale

4 artificial bristle
Kunstborste (f)
cerda (f) artificial
crin (m) artificiel
setola (f) artificiale

5 artificial daylight
künstliches Tageslicht (n)
luz (f) diurna artificial
lumière (f) artificielle
luce (f) artificiale

6 artificial fibre
Chemiefaser (f), Kunstfaser (f)
fibra (f) artificial
fibre (f) artificielle
fibra (f) artificiale

7 artificial fur
Pelzimitation (f)
piel (f) artificial
fourrure (f) synthétique
pelliccia (f) artificiale

8 artificial silk
Kunsteide (f)
seda (f) artificial
soie (f) artificielle
seta (f) artificiale

9 asbestos
Asbest (m)
amianto (m)
amiante (m)
amianto (m)

10 asbestos fabric
Asbestgewebe (n pl)
tejidos (m pl) de amianto
tissus (m pl) en amiante
tessuti (m pl) di amianto

11 asbestos fibre
Asbestfaser (f)
fibra (f) de amianto
fibre (f) d'amiante
fibra (f) di amianto

12 asbestos yarns
Asbestgarne (n pl)
hilados (m pl) de amianto
filés (m pl) en amiante
filati (m pl) di amianto

13 ascertain, verify (v)
feststellen
constatar, verificar
constater, vérifier
accertare, verificare

14 asking price
Forderungspreis (m)
precio (m) de oferta
prix (m) demandé
prezzo (m) di offerta

15 assemble, set up (v)
montieren, zusammensetzen
montar
monter
montare

16 assistant
Hilfsmittel (n)
auxiliar (m)
auxiliaire (m), adjuvant (m)
ausiliario (m)

17 astrakhan
Astrachan
astracán (m)
astrakan (m)
astrakan (m)

18 atmosphere
Atmosphäre (f)
atmósfera (f)
atmosphère (f)
atmosfera (f)

1 atmosphere for testing
Prüfatmosphäre (f)
condiciones (f pl) de prueba
atmosphère (f) d'essai (m)
condizioni per le prove

2 atmospheric pressure
Atmosphärendruck (m)
presión (f) atmosférica
pression (f) atmosphérique
pressione (f) atmosferica

3 atomizer
Zerstäuber (m)
atomizador
atomiseur (m)
nebulizzatore (m)

4 attack; attach, fix (v)
ätzen, angreifen; anheften
atacar; fijar
attaquer, attacher
attaccare

5 attempt, trial
Versuch (m)
prueba (f), ensayo (m)
essai
tentativo (m), prova (f)

6 auburn
kastanienbraun (adj)
color (m) castaño
auburn (adj inv)
castano dorato

7 auction
Auktion (f)
subasta (f)
vente (f) aux enchères
asta (f)

8 auctioneer
Auktionator (m)
subastador (m), subastadora (f)
commissaire-priseur (m)
banditore (m)

9 audit
Revision (f)
revisión (f) de cuentas
vérification (f) comptable
revisione (f)

10 autoclave
Autoklav (m)
autoclave (m)
autoclave (m)
autoclave (f)

11 auto-leveller
Regelstrecke (f)
nivelador (m) automático
autorégulateur
livellatore (m) automatico

12 automated control systems
automatisierte Steueranlagen (f pl)
sistemas (m pl) de control automático
systèmes (m pl) de commande automatique
sistemi (m pl) di controllo automatizzato

13 automatic
automatisch (adj)
automático
automatique
automatico

14 automatic screen printing
automatischer Filmdruck (m)
estampado (m) automático con plantilla tamiz
impression (f) à cadre automatique
stampa (f) automatica a quadri

15 automatic shuttle change box
automatischer Schützenwechsel (m)
cambio (m) de lanzaderas automático
boîte (f) de changement (m) de navette (f) automatique
cambio (m) della navetta automatico, arralgimento (f)

16 automatic winding
Spulautomat (m)
bobinado (m) automático
bobinage (m) automatique
incannatura (f) automatica

17 automation
Automatisierung (f)
automatización (f)
automation (f)
automazione (f)

18 auxiliary
Hilfsmittel (n)
auxiliar
auxiliaire
ausiliario (agg)

1 auxiliary products
Hilfsprodukte (n pl)
productos (m pl) auxiliares
produits (m pl) auxiliaires
prodotti (m pl) ausiliari

2 available
verfügbar
disponible
disponible
disponibile

3 average
Durchschnitt (m)
promedio (m)
moyenne (f)
medio

4 avoid, prevent (v)
umgehen, verhüten, verhindern
evitar, impedir
éviter, empêcher
evitare, impedire

5 awning
Markise (f)
toldo (m)
store (m)
tenda (f)

6 Axminster
Axminsterteppich (m)
Axminster (m)
axminster
tappeto (m) Axminster

7 azine dye
Azinfarbstoff (m)
colorante (m) azino
azine
colorante (m) azinico

8 azo dye
Azofarbstoff (m)
colorante (m) azoico
colorant (m) azoïque
azocolorante (m)

9 azure
azurblau (adj)
azul celeste
azur
azzurro

B

10 baby blue
Baby Blau (adj)
azul claro
bleu tendre
celeste chiaro

11 backlog
Arbeitsrückstand (m)
acumulación (f)
retard (m) de travail
scorta (f), riserva (f), accumulo (m)

12 back-part [loom]
hinterer Abschnitt (m)
parte (f) trasera del telar
derrière (m) [d'un métier]
parte (f) posteriore [telaio]

13 back-washing
lissieren (m)
relavar (m)
lissage (m)
lavaggio (m) a controcorrente

14 bad debt
schlechter Außenstand (m)
débito (m) no liquidado
mauvaise créance (f)
credito (m) inesigibile

15 badges
Abzeichen (n pl)
insignias (f pl)
insignes (m pl)
distintivi (m pl)

16 bag
Sack (m)
saco (m)
sac (m)
sacco (m)

17 bagging [cloth]
Sackleinwan (f)
arpillera (f)
toile (f) à sac
tela (f) da sacchi

1 balance of trade
Handelsbilanz (f)
balance comercial (m)
balance (f) commerciate
bilancia (f) del commercio

2 balances
Ausgleich (m) Bilantz
balanzas (f pl)
balances (f pl)
bilance (f pl)

3 bald
kahl (adj)
calvo
lisse
liscio, calvo, nudo

4 bale
Ballen (m)
fardo (m), bala (f)
balle (f)
balla (f)

5 bale-breaker
Ballenöffner (m)
abridor (m) de fardos
ouvreuse (f) de balles
apriballe (m)

6 bale-strapping
Ballenbänder (n pl)
correaje (m) de pacas
cerclage (m) de balles
cinghie (f) per balle

7 baling
Verpacken in Ballen
embalar (m)
mise (f) en balles
imballaggio (m)

8 ballers, ball winders
Wickelmaschine (f), Knäuelmaschine (f)
apelotonadores (m pl)
pelotonneuses (f pl)
imballatrice (f pl), gomitolatrice

9 ball-warps of the mop industry
Kettenwickel für die Mopindustrie
tramas (f pl) redondas para la fabricación de aljofifas
chaîne (f) ourdie en pelotes pour l'industrie des balais à franges
orditoi (m pl) in gomitoli per l'industria di raclazza

10 bandages – crêpe
Verbandsstoffe (m pl)
vendas (f pl) de crespones
bandes (f pl) velpeau
bende (f pl) – crepe

11 banging off [loom]
Abschlagen (m)
arrancar (m) la lanzadera
blocage (m) du métier à tisser
impiccarsi [telaio] (v intr)

12 bank
Bank (f)
banco (m)
banque (f)
banca (f)

13 bank account
Bankkonto (n)
cuenta (f) bancaria
compte (m) en banque
conto (m) bancario

14 bank base rate
Eckzinssatz (m)
cambio (m) de base
taux (m) de base bancaire
saggio (m) base bancario

15 bank clearing
Abrechnung (f)
banco (m) de liquidación
clearing (m) en banque
titolo (m) di compensazione

16 bank holiday
Bankfeiertag (m)
fiesta (f) nacional
fête (f) légale, jour Férié
festa (f) civile, nazionale

17 bank manager
Filialleiter (m)
director (m) de banco
directeur (m) de banque
direttore (m) di banca

18 bank note
Banknote (f)
billete (m) de banco
billet (m) de banque
banconota (f), biglietto (m) di banca

1 bank rate
Diskontsatz (m)
cambio (m) oficial
taux (m) d'escompte officiel
tasso (m) ufficiale di sconto

2 bank statement
Kontoauszug (m)
balance (m) de banco
relevé (m) de compte
rendiconto (m) bancario

3 banker
Bankier (m)
banquero (m)
banquier (m)
banchiere (m)

4 banker's order
Banküberweisungsuftrag (m)
orden (m) bancaria
ordre (m) de virement bancaire
ordine (m) di banca, ordine (m) di pagamento

5 bankrupt
bankrott (adj)
bancarrota (f)
failli
bancarotta (f)

6 Barathea weave
Barathea Gewebe (n)
tejido (m) baratea
armure (f) barathéa
tessuto (m) incrociato con disegni in relievi

7 Barbary sheep
Barbary Schaf (n)
oveja (f) norteafricana
mouton (m) de Barbarie
pecora (f) africana

8 bargain
güstiges Angebot (n)
ganga (f)
bon marché (m)
affare (m), occasione (f)

9 barleycorn weave
Gerstenkorngewebe (n)
tejido (m) de cebada
armure (f) grain d'orge
tessuto (m)/armatura (f) a chicco d'orzo

10 barriness [weft]
Schußstreifigkeit
barrado (m) en la dirección de la trama
barrures (f pl) en trame
barriness [trama]/barrattare di trama

11 barter
Tausch (m)
regatear
échange (f)
baratto (m), scambio (m)

12 basket weave
Korbgewebe (n), Panamagewebe (n)
tejido (m) de cesta
armure (f) Panama
armatura (f) panama

13 batch drying
Partietrocknen (n)
secar (m) por grupos
séchage (m) par lot
essiccazione (f) discontinua, essiccamento partita

14 batch-dyeing
Partiefärben (n)
teñir por lotes
teinture (f) par lot
tintura (f) discontinua, tintura di partita

15 batching
Batschen (m)
combinar (m)
enroulement (m) de la chaîne
oliaggio (m)

16 batching machinery
Batschenmaschine (f)
maquinaria (f) de combinar
équipement (m) à enrouler
macchinario (m) da oliaggio

17 bath
Bad (n)
baño (m)
bain (m)
bagno (m)

18 bath-dyeing
Färbebad (n)
teñir (m) en baño de colorante
teinture (f) en bain
tintura (f) in bagno

1 battery
Magazin (n)
pila (f)
barillet (m) chargeur
batteria (f)

2 beam cover
Baumabdeckung (f)
cubierta (f) del plegador
plateau (m) d'ensouple
copri-subbio (m)

3 beam flange
Baumscheibe (f)
reborde (m) del plegador
disque (m) de l'ensouple
disco (m) del subbio

4 beam [warp]
Kettbaum (m)
enjulio (m)
ensouple (m) de derrière
cilindri (m pl) [orditura a], cubbia d'ordito

5 beaming
Bäumen (n)
devanar (m) en el plegador
ensouplage (m)
montaggio (m), insubbiamento

6 beaming machine
Bäumvorrichtung (f)
máquina (f) de devanar en enjulio
machine (f) d'ensouplage
macchina (f) da montaggio, macchina d'insubbiamento

7 bearing
Lager (n)
cojinete (m)
coussinet (m)
sopporto (m), cuscinetto (m)

8 beating up [loom]
Aufschlagen (n)
peinar (m) la trama
battage (m)
unione (f) della trama inserita al tessuto

9 bedding
Bettzeug (n)
ropa (f) de cama
draps (m pl) et taies (f pl)
biancheria (f) da letto

10 bedding – cotton
Bettwäsche – Baumwolle (f)
ropa (f) de cama de algodón
draps (m pl) et taies (f pl) en coton
biancheria (f) da letto – cotone

11 bedding – household textile
Haushalttextilien (n pl)
ropa (f) de cama – tejidos de casa
literie (f) – linge (m) de maison
biancheria (f) da letto te ssuti da casa

12 bedding – polycotton
Bettwäsche – Polyester/Baumwolle (f)
ropa (f) de cama de poliester/algodón
draps (m pl) et taies (f pl) en polyester/coton
biancheria (f) e coperte da letto – poliestere/cotone

13 beige
beige (adj)
beige
beige
beige

14 belt
Riemen (m)
correa (f)
courroie (f)
cinghia (f)

15 belt drive
Riemenantrieb (m)
transmisión (f) por correa
commande (f) par courroie
cinghia di trasmissione

16 belting
Triebriemen (m)
correa (f)
courroies (f pl)
materiale (m) per cinghie

17 benzene
Benzol (n)
benceno (m)
benzène (m)
benzolo (m)

1 bill
Rechnung (f)
documento (m)
effet (m)
cambiale (f), conto (m)

2 bill of exchange
Wechsel (m)
letra (f) de cambio
lettre (f) de change
tratta (f), vaglia (m) cambiario, cambiale

3 bill of lading
Frachtbrief (m)
conocimiento (m) de embarque
connaissement (m)
polizza (f) di carico

4 bill of sale
Kaufbrief (m)
escritura (f) de venta
acte (f) de vente
atto (m) di cessione, atto di vendita

5 bird's-eye weave
Pfauenaugengewebe (n)
tejido (m) moteado
armure (f) oeil de perdrix
tessuto (m)/armatura (f) ad occhio di pavone

6 black
schwarz (adj)
negro
noir
nero

7 black-faced sheep
schwarzköpfiges Schaf (n)
oveja (f) de cara negra
mouton (m) à tête noire
pecora (f) dal muso nero

8 black sheep
schwarzes Schaf (n)
oveja (f) negra
brebis (f) noire
pecora (f) nera

9 blanket [cloth]
Decke (f)
manta (f)
couverture (f), housse (f)
coperta (f)

10 blanket manufacturers
Deckenhersteller (m pl)
fabricantes (m pl) de mantas
fabricants (m pl) de couvertures
produttori (m pl) di coperte

11 bleachers
Bleicher (m pl)
blanqueadores (m pl)
blanchisseurs (m pl)
sbiancanti (m), imbiancatori

12 bleachers – cloth
Bleicher (m pl) – Tuch, Tuchbleicher (m pl)
blanqueadores (m pl) de telas
blanchisseurs (m pl) de tissu
recipienti (m pl) per candeggiare – tessuto

13 bleachers – cotton
Baumwollbleicher (m pl)
blanqueadores (m pl) de algodón
blanchisseurs (m pl) de coton
imbiancantori di cotone

14 bleachers – cotton yarns
Baumwollgarnbleicher (m pl)
blanqueadores (m pl) de hilos de algodón
blanchisseurs (m pl) de fils de coton
imbiancatori di filati di cotone

15 bleachers – fabric – cotton
Baumwollgewebebleicher (m pl)
blanqueadores (m pl) de telas de algodón
blanchisseurs (m pl) de tissu de coton
imbiancatori di tessuto-cotone

16 bleachers – fabric – cotton/polyester
Baumwoll/Polyestergewebebleicher (m pl)
blanqueadores (m pl) de telas de algodón/poliéster
blanchisseurs (m pl) de polyester/coton
imbiancatori di tessuto-cotone/ poliestere

17 bleachers – fabric – viscose
Viskosegewebebleicher (m pl)
blanqueadores (m pl) de telas de viscosa
blanchisseurs (m pl) de viscose
imbiancatori di tessuto-viscosa

1 bleachers – fabric – viscose/flax
Viskose/Flachsgewebebleicher (m pl)
blanqueadores (m pl) de telas de viscosa/lino
blanchisseurs (m pl) de viscose/lin
imbiancatori di tessuto viscosa fibra di lino

2 bleachers – fabrics – viscose/tejidos
Viskose/Polyestergewebebleicher (m pl)
blanqueadores (m pl) de telas de viscosa/poliéster
blanchisseurs (m pl) de viscose/polyester
imbiancatori di tessuto – viscosa/poliestere

3 bleachers – fabric – viscose/polyester & flax
Viskose/Polyester und Flachsgewebebleicher (m pl)
blanqueadores (m pl) de tejidos de viscosa, poliéster y lino
blanchisseurs (m pl) de viscose/polyester et lin
imbiancatori di tessuto – viscosa/poliestere e fibra di lino

4 bleachers – hank
Gebindbleicher (m pl), Strangbleicher (m pl)
blanqueadores (m pl) de madejas
blanchisseurs (m pl) d'écheveaux
imbiancatori di matasse

5 bleachers – loose fibre
Bleicher (m pl) für lose Fasern
blanqueadores (m pl) de fibras sueltas
blanchisseurs (m pl) de fibres en bourre
imbiancatori di fibre sciolte

6 bleachers – loose wool
Bleicher (m pl) für lose Wolle
blanqueadores (m pl) de lana suelta
blanchisseurs (m pl) de laine en bourre
imbiancatori di lana sciolta

7 bleachers – silk
Bleicher (m pl) – Seide, Bleicher (m pl) für Seide
blanqueadores (m pl) de seda
blanchisseurs (m pl) de soie
imbiancatori di seta

8 bleachers – synthetic yarns
Bleicher (m pl) – Synthetikgarne, Bleicher (m pl) für Synthetikgarne
blanqueadores (m pl) de hilos sintéticos
blanchisseurs (m pl) de fils synthétiques
imbiancatori di filati sintetici

9 bleachers – tops
Kammzugbleicher (m pl)
blanqueadores (m pl) de hilados
blanchisseurs (m pl) de rubans
imbiancatori di nastri di lana pettinata

10 bleachers – yarn
Garnbleicher (m pl)
blanqueadores (m pl) de hilos
blanchisseurs (m pl) de fil
imbiancatori di filati

11 bleaches
Bleichmittel (n pl)
blanquimientos (m pl)
produits (m pl) blanchissants
candeggine (f pl)

12 bleaching
Bleichen (n)
blanquear (m)
blanchiment (m)
candeggio (m), sbiancanti (m pl)

13 bleaching agents
Bleichmittel (n pl)
agentes (m pl) de blanquear
agents (m pl) de blanchiment
candeggianti (m pl), sbiancanti (m pl)

14 blenders
Mischer (m pl)
mezcladores (m pl)
mélangeurs (m pl)
miscelatori (m pl), mischiatrici (f pl)

15 blenders – loose synthetic
Mischer (m pl) für lose Synthetikgewebe
mezcladores (m pl) de sintético suelto
mélangeurs (m pl) de synthétiques en bourre
mescolatrici (f pl) – fibre sintetiche sciolte

1 blenders – loose waste
Mischer (m pl) für losen Abfall
mezcladores (m pl) de residuos sueltos
mélangeurs (m pl) de déchets en bourre
miscelatori (m pl), mischiatori (m pl) – cascami scioti

2 blenders – loose wool
Mischer (m pl) für lose Wolle
mezcladores (m pl) de lana suelta
mélangeurs (m pl) de laine en bourre
miscelatori (m pl), mischiatori (m pl) – lana sciolta

3 blenders – wool and hair
Mischer (m pl) – Wolle und Haar
mezcladores (m pl) de lana y pelo
mélangeurs (m pl) de laine et poils
miscelatori (m pl), mischiatori (m pl) – lana e pelo

4 blenders and oilers
Mischer (m pl) und Schmälter (m pl)
mezcladores (m pl) y lubricantes
mélangeurs (m pl) et huileurs (m pl)
miscelatori (m pl), mischiatori (m pl) e oliatori (m pl)

5 blenders and oilers – loose synthetic
Mischer (m pl) und Schmätzer (m pl) für lose Synthetikgewebe
mezcladores (m pl) y lubricantes (m pl) de sintético suelto
mélangeurs (m pl) et huileurs (m pl) de synthétiques en bourre
miscelatori (m pl), miscelatrici (f pl) e oliatori (m pl) – fibre sintetiche sciolte

6 blenders and oilers – loose waste
Mischer (m pl) und Schmätzer (m pl) für losen Abfall
mezcladores (m pl) y lubricantes de residuos sueltos
mélangeurs (m pl) et huileurs (m pl) de déchets en bourre
miscelatori (m pl), miscelatrici (f pl) e oliatori – cascame sciolto

7 blenders and oilers – loose wool
Mischer (m pl) und Schmätzer (m pl) für lose Wolle
mezcladores (m pl) y aceitadores de lana suelta
mélangeurs (m pl) et huileurs (m pl) de laine en bourre
miscelatori (m pl), miscelatrici (f pl) e oliatori – lana sciolta

8 blending
vermischen
mezclar (m)
mélange (f)
mischia (f)

9 blending oils
Mischöle (n pl)
aceites para la mezcla lubricantes
huiles (f pl) de mélange
oli (m pl) per la mischia

10 blending oils – free scouring mineral oils
Mischöle – auswaschbare Mineralöle
aceites (f pl) para mezclarse – aceites minerales para desengrasarse fácilmente
huiles (f pl) de mélange – huiles minérales de lavage
oli (m pl) per la mischia – oli minerali privi di scorie

11 blending oils – lubricants
Mischöle Gleitmittel (m pl)
aceites para la mezcla lubricantes
huiles (f pl) de mélange – lubrifiants (m pl)
oli (m pl) per la mischia – lubrificanti

12 blending oils – oleines
Mischöle – Oleine (m pl)
aceites para la mezcla y oléínas
huiles (f pl) de mélange – oléines
oli (m pl) per la mischia – oleine

13 blonde
blond (adj)
rubio
blond
biondo

14 blood-red
blutrot (adj)
color (m) de sangre
rouge sang
rosso sangue

15 blow blending
Blasmischungen (f pl)
mezcla por medio de soplado
mélange (m) par soufflage
mischia (f) mediante soffiatura

1 blowing
Abblasen (m)
soplar (m)
soufflage (m)
soffiamento

2 blown finish
Gebläseappretur (f)
acabado por soplado
apprêt (m) gonflant
finissaggio soffiata

3 blue
blau (adj)
azul
bleu
blu

4 blue-black
blauschwarz (adj)
negro azulado
bleu-noir
blu nero

5 board meeting
Aufsichtsratsitzung (f)
reunión (f) del consejo de administración
réunion de conseil d'administration
assemblea (f) del consiglio di amministrazione

6 boardroom
Sitzungszimmer (n)
sala (f) de reuniones del consejo de administración
salle de réunion de conseil d'administration
sala (f) del consiglio di amministrazione

7 bobbin
Spule (f)
bobina
bobine (f), canette (f)
bobina (f), rocchetto (m), spola (f)

8 bobbin stripping
Spulenreinigung (f)
deshilado de los husos (m)
nettoyage (m) des canettes
ripulitura (f) della bobina

9 boil (v)
sieden
hervir
bouillir
bollire

10 boiler
Sieden (m)
caldera (f)
chaudière (f)
bollitore (m), caldaia (f)

11 botany wool
Botanywolle (f)
lana (f) de botany
laine (f) mérinos
lana (f) australiana

12 bottle green
flaschengrün (adj)
verde botella
vert bouteille
verde bottiglia

13 bottom shed [loom]
Tieffach (n)
taller (m) inferior
foule (f) de rabat
passo (m) d'ordito inferiore [telaio]

14 bouché weave
Bouclégewebe (n)
tejer (m) de astracán
armure (f) bouclée
tessuto (m) boucle

15 bowed cloth
Verzogenes Gewebe (n)
tela (f) deformada
tissu (m) arqué
tessuto (m) currato

16 bowing
Verkrümmen des Gewebes, Verziehen (n) oder Verzerren (n)
deformación (f)
bombage (m)
archeggio (m), flessione (f)

17 box-loom
Wechselstuhl (m)
tejer (m) de caja
métier (m) à plusieurs navettes
telaio (m) a più cassette

18 Bradford conditioning house
Bradforder Konditionieranstalt (f)
casa (f) de acondicionar de Bradford
condition (f) publique textile Bradford
stablimento (m) di condizionatura di Bradford

1 braiding machine
Litzennähmaschine (f)
máquina (f) de trenzar
méticr (m) à tresser
intrecciatrice

2 braids
Borten (f pl)
trenzas (f pl)
tresses (f pl)
galloni (m pl), bordure (f pl), passamani (m pl)

3 break down
Unterbrechung (f), Panne (f)
avería (f)
panne (f)
rottura (f), guasto (m)

4 breaking strength
Reißfestigkeit (f)
resistencia (f) a la roturda
résistance à la cassure
resistenza (f) alla rottura

5 breast beam [loom]
Brustbaum (m)
enjulio (m) de pecho
ensouple (m) de devant
subbio (m) anteriore, pettorale [telaio]

6 brick red
ziegelrot (adj)
rojo anaranjado
rouge brique
rosso mattone

7 bright
hell (adj)
claro
clair
vivace, brillante

8 British Summer Time
britische Sommerzeit (f)
hora de verano inglesa
heure (f) d'été britannique
ora (f) legale inglese

9 British wool
britische Wolle (f)
lana inglesa (f)
laine (f) britannique
lana (f) inglese

10 broad loom
Breitwebstuhl (m)
telar (m) ancho
métier m) à tisser large
telaio (m) largo

11 broad width
Doppelbreite (f)
anchura (f) ancha
grande largeur (f)
doppia carglezza (f) altezza larga

12 brocade [cloth]
Brokat (m)
brocado (m)
brocart (m)
broccato (m)

13 brown
braun
marrón
marron, brun
marrone

14 brush
Bürste (f)
cepillo (m)
brosse (f)
spazzola (f)

15 brushes
Bürsten (f pl)
cepillos (m pl)
brosses (f pl)
spazzole (f pl)

16 brushes – all types of twisted in wire tubes – pneumatic etc.
Bürsten (f pl) – alle Arten von eingedrehten Rutendüsen – Pneumatik usw.
cepillos (m pl) de toda clase, de alambres, torcidos tubos – neumáticos etc.
brosses (f pl) – toutes sortes, tordues dans tubes métalliques – pneumatiques, etc.
spazzole (f pl) – tutti i tipi di ritori in tubi di fili – pneumatici, ecc.

17 brushes – carpet trade
Bürsten für den Teppichhandel
cepillos (m pl) para la industria de alfombras
brosses (f pl) pour l'industrie des tapis
spazzole (f pl) – per tappeti

1 brushes – combing
Bürsten (f pl) – zum Kämmen
cepillos (m pl) de fibra carta desechables
brosses (f pl) de peignage
spazzole (f pl) – pettinatura

2 brushes – disposable French noils
Bürsten (f pl) – französsiche Einweg-Kämmlinge
cepillos (m pl) disponibles de fibras cortas
brosses – blousses française à jeter
spazzole (f pl) – cascame di pettinatura francese da gettare dopo l'uso

3 brushes – dyeing and finishing
Bürsten (f pl) – zum Färben und Appretieren
cepillos (m pl) para teñir y acabar
brosses (f pl) – teinture et finissage
spazzole (f pl) – tintura e finissaggio

4 brushes – industrial
Bürsten (f pl) – für die Industrie
cepillos (m pl) industriales
brosses (f pl) industrielles
spazzole (f pl) – industriali

5 brushes – textile manufactured and repaired
Bürsten (f pl) – Textil gefertigt und repariert
cepillos para tejidos – fabricados y reparados
brosses (f) textiles fabriquées et réparées
spazzole (f pl) – produttori e riparatori – settore tessile

6 brushing
bürsten
cepillar (m)
brossage (m)
spazzolatura

7 buff [colour]
lederfarben (adj)
color (m) de ante
jaune clair
fulvo [colore]

8 building contractors
Bauunternehmer (m pl)
contratistas (m pl) de obras de edificios
entrepreneurs (m pl) de bâtiments
imprenditori (m pl) di lavori

9 building maintenance
Gebäudewartung (f)
mantenimento de edificios
entretien (m) de bâtiments
manutenzione (f) di edifici, mantenimento d'edifici

10 bulking
Bauschen (n)
embalar (m) en bulto/volumen (m)
traitement (m) de gonflant
rigonfiamento (m), imbottitura (f), voluminizzare

11 bunch maker
Bündler (m)
fabricante (m) de haces de fibras para comparar colores
fabricant (m) d'échevettes
dispositivo (m) per creare fiocchi

12 bunch makers
Bündler (m pl)
fabricantes (m pl) de haces de fibras para comparar colores
fabricants (m pl) d'échevettes
dispositivi (m pl) per creare fiocchi

13 bunting [cloth]
Fahnentuch (n)
tejidos para banderas
molleton (m) à drapeaux
stamigna (f)

14 burl
Noppe (f)
desmotar (m)
nope (f)
pulire i nodi

15 burling irons
Nappeisen (n pl)
desmotaderas (f pl)
pinces (f pl) à noper
pinzo per la pulitura di nodi

16 burling [cloth]
Nappenstoff (m)
desmotar (f) [tela]
nopage (m)
eliminazione dei nodi [tessuto]

17 burning
Ausbrennen (n)
quemando
brûlage
bruciatura

1 burnt-orange
gebrantes orange
naranja (f) quemada
rouge orange
color arancione bruciato

2 burr
Klette (f)
mota (f)
chardon (m)
lappola (f)

3 burry wool
Klettwolle (f)
lana (f) con motas
laine (f) chardonneuse
lana (f) lappolosa

4 bursting point
Berstpunkt (m)
punto (m) de reventamiento
point (m) d'éclatement
limite (m) di sopportazione, prova di resistenza allo scoppio (m)

5 butter-muslin [cloth]
Musselingaze
percal (m)
étamine (f)
mussola (f) marrone

6 buyer
Käufer (m)
comprador (m)
acheteur (m)
compratore (m)

7 buyers' market
Käufermarkt (m)
mercado (m) de compradores
marché (m) demandeur
mercato (m) del compratore, mercato al ribasso

8 buying
Kaufen (n)
comprar (m)
achat (m)
compra/approvvigionamento

9 by-product
Nebenprodukt (n)
derivado (m)
sous-produit (m)
prodotto secondario (m)

C

10 cable yarn
Cablégarn (m)
hilado (m) cableado
fil (m) câblé
filato (m) cablé

11 cake
Spinnkuchen (m)
corona (f), rodete (m)
gâteau (m)
torta (f)

12 cake-dyed
spinnkuchengefärbt (adj)
teñido en corona
teint en gâteau
tinto in focaccia

13 calcium
Kalzium (n), Calcium (n)
calcio (m)
calcium (m)
calcio (m)

14 calcium carbonate
Kalziumkarbonat (n)
carbonato (m) de calcio
carbonate (m) de calcium
carbonato (m) di calcio

15 calcium chloride
Kalziumchlorid (n), Calciumchlorid (n), Chlorkalzium (n)
clórido (m) de calcio, cloruro (m) de calcio
chlorure (m) de calcium
cloruro (m) di calcio

16 calcium nitrate
Kalziumnitrat (n)
nitrato (m) cálcico
nitrate (m) de calcium
nitrato (m) di calcio

17 calcium phosphate
Kalziumphosphat (n)
fosfato (m) de calco
phosphate (m) de chaux
fosfato (m) di calcio

1 calcium sulphate
Kalziumsulfat (n), Calciumsulfat (n)
sulfato (m) de calcio, sulfato (m) cálcico
sulfate (m) de calcium
solfato (m) di calcio

2 calculate (v)
berechnen, kalkulieren
calcular
calculer
calcolare

3 calculation
Berechnung (f), Kalkulation (f)
cálculo (m)
calcul (m)
calcolo (m)

4 calendar
Kalander (m)
calandra (f), calandria (f)
calandre (f)
calandra (f)

5 calendering
Kalandrieren (n), Kalandern (n), Kalandrierung (f)
aprensadores (m pl), calandrado (m)
calandrage (m)
calandratura (f)

6 calibration
Eichung (f)
contraste (m)
tarage (m)
taratura (f)

7 calorific value
Heizwert (m)
potencia (f) calorífica
pouvoir (m) calorifique
potere (m) calorifico

8 camel [colour]
kamelfarben (adj)
color camello
chameau [couleur]
cammello [color]

9 canary yellow
kanariengelb (adj)
amarillo canario
jaune canari
giallo canarino

10 canvas [cloth]
Leinwand (f), Kanevas (m)
lona (f)
grosse toile (f), canevas (m)
tela (f) olona, canovaccio (m)

11 capacity
Kapazität (f), Leistung (f), Potenz (f)
capacidad (f), potencia (f), energía (f)
capacité (f), puissance (f)
capacità (f), potenza (f)

12 capillary
kapillar (adj)
capilar
capillaire
capillare

13 cap-spinning frame
Glockenspinnmaschine (f)
continua (f) de hilar de campanas, continua (f) de hilar de capacetes
métier (m) à filer à cloches, continu (m) à cloche
filatoio (m) a campana

14 cap-twisting frame
Glockenzwirnmaschine (f)
retorcedora (f) de caperuzas
moulineuse (f) à cloches
torcitoio (m) a campana

15 carbohydrate
Kohlenhydrat (n)
hidrato (m) de carbono
glucide (m)
carboidrato

16 carbon dioxide
Kohlendioxyd (n)
anhídrido (m) carbónico, dióxido (m) de carbono
dioxyde (m) de carbone, bioxyde (m) de carbone
anidride (f) carbonica, biossido (m) di carbonio

17 carbonic acid
Kohlensäure (f)
ácido (m) carbónico
acide (m) carbonique
acido (m) carbonico

18 carbonization
Karbonisation (f), Auskohlen (n)
carbonización (f)
carbonisage (m)
carbonizzazione (f)

1 carbonize (v)
karbonisieren
carbonizar
carboniser
carbonizzare

2 carbonized wool
karbonisierte Wolle (f)
lana (f) carbonizada
laine (f) carbonisée
lana (f) carbonizzata

3 carbonizers – cloth
Tuchkarbonisierer (m pl)
carbonizadores (m pl) de telas
carboniseurs (m pl) de tissu
carbonizzatori (m pl) – tessuto

4 carbonizers – rag
Lumpenkarbonisierer (m pl)
carbonizadores (m pl) de trapos
carboniseurs (m pl) de déchets
carbonizzatori (m pl) – stracci

5 carbonizers – waste
Abfallkarbonisierer (m pl)
carbonizadores (m pl) de residuos
carboniseurs (m pl) de déchets
carbonizzatori (m pl) – cascame

6 carbonizers – wool
Wollkarbonisierer (m pl)
carbonizadores (m pl) de lana
carboniseurs (m pl) de laine
carbonizzatori (m pl) – lana

7 carbonizing
Verkohlen (n), Karbonisation (f), Auskohlen (n)
carbonizar (m), carbonización
carbonisation (f), carbonisage (m)
carbonizzazione (f)

8 carbon monoxide
Kohlenoxyd (n), Kohlenmonoxide (n pl)
monóxido de carbono, óxido (m) de carbono
oxyde (m) de carbone
ossido (m) di carbonio

9 carbon tetrachloride
Tetrachlorkohlenstoff (m)
tetracloruro (m) de carbono
tétrachlorure (m) de carbone
tetracloruro (m) di carbonio

10 card
Karde (f), Krempel (f)
carda (f)
carde (f)
carda (f)

11 card (v)
kardieren, kratzen, krempeln
cardar
carder
cardare

12 cardboard
Karton (m)
cartón (m)
carton (m)
cartone (m)

13 cardboard boxes and cartons
Pappkartons und Kisten
cajas (f pl) de cartón
caisses (f pl) et boîtes (f pl) en carton
scatole (f pl) di cartone, scatole

14 card clothing
Kardenbeschlag (m), Kardengarnitur (f), Krempelbeschlag (m), Kratzenbeschlag (m)
tela (f) de cardas, guarnición de cardas
garniture (f) de carde
guarnizione (f) di carda, scardasso (m)

15 carded
kardiert (adj)
cardado
cardé
cardato

16 carders
Krempler (m pl)
cardadores (m pl)
cardeuses (f pl)
cardatrici (f pl)

17 card-grinding
Kardenschleifen (n)
afilado (m) de cardas
aiguisage (m) des cardes
molatura (f) delle guarizioni di carda

18 carded cotton
Streichbaumwolle (f), kardierte Baumwolle (f)
algodón (m) cardado
coton (m) cardé
cotone (m) cardato

1 carded wool
Streichwolle (f)
lana (f) cardada
laine cardée
lana (f) cardata

2 carded yarn
Streichgarn (n)
hilo (m) cardado, hilado (m) cardado
fil (m) cardé
filato (m) cardato

3 carding
Kardieren (n), Krempeln (n)
cardón (m), cardado (m), cardadura (f)
cardage (m)
cardatura (f)

4 carding engine
Kardiermaschine (f)
motor (m) de cardón
machine (f) à carder
motori di carda

5 carding nails
Kardiernägel (m pl)
clavos (m pl) de cardar
clous (m pl) de la carde
punte (f pl) per carda

6 card-lacing machine
Kartenbindemaschine (f)
máquina (f) de coser los cartones jacquard
machine (f) à nouer les cartons
cucitrice (f) per cartoni jacquard

7 card-punching machine
Kartenstanze (f)
máquina (f) perforadora de cartones
matrice (f) à cartons
punzonatrice (f) di cartoni

8 card room
Kardierraum (m), Karderie (f), Kardensaal (m), Krempelsaal (m)
sala (f) de cardas
carderie (f), salle (f) de cardage, salle (f) de cardes
carderia (f), sala (f) delle carde

9 card set
Krempelsatz
equipo (m) para el cardado
réglage (m) de cardes
serie (f) di carde, assortimento di carde

10 card silver
Kardenband (n)
cinta (f) de carda
ruban (m) de carde
nastro (m) di carda

11 card waste
Kardenabfall (m)
desperdicios (m pl) de carda
déchets (m pl) de carde
cascame (m) di carda

12 card wire
Kratzendraht (m)
alambre (m) de cardas, alambre (m) de guarnición de carda
fil (m) de garniture, fil (m) pour garniture de carde
dente (m) di carda, filo (m) per guarnizione di carda

13 carpet
Teppich (m)
alfombra
tapis (m)
tappeto (m)

14 carpet and rug manufacturers
Teppich- und Vorlegerhersteller (m pl)
fabricantes (m pl) de alfombras y alfombrillas
fabricants (m pl) de moquette et tapis
produttori (m pl) di tappeti e tappetini

15 carpet backing
Teppichgrund (m)
tejido (m) de fondo para alfombras
fond (m) du tapis
fondo (m) del tappeto

16 carpet loom
Teppichwebstuhl (m)
telar (m) para alfombras
métier à tapis
telaio (m) per tappeti

17 carpet shearing machine
Teppichschermaschine (f)
tundidora (f) de alfombras
tondeuse (f) pour tapis
climatrice (f) per tappeti

18 carpet weaving
Teppichweberei (f)
tejedura de alfombras
tissage (m) de tapis
tessitura (f) dei tappeti

1 carriage
Wagen (m)
carro (m)
chariot (m)
carro (m)

2 carriage-free
frachtfrei (adj)
sin gastos de transporte
franc de port
franco di porto

3 carrier
Färbebeschleuniger (m), Beschleuniger (m)
acelerador (m) [de tintura]
accélerateur (m) [de teinture]
accelerante (m) di tintura, carrier (m)

4 carton
Karton (m), Gehäuse (m), Schachtel (f)
caja (f)
boîte (f), carter (m)
scatola (f)

5 case
Kasten (m)
caja (f)
caisse (f)
cassa (f)

6 casein fibre
Kaseinfaser (f)
fibra (f) de caseína
fibre (f) de caséine
fibra (f) caseinica

7 cash-flow
Geldfluß (m)
movimiento (m) de capital
marge (f) brute d'autofinancement
movimento di capitale

8 cashmere
Kaschmir (m)
cachemir (m)
cachemire (m)
cashmere (m)

9 castors and wheels
Rollen (f pl) und Räder (n pl)
ruedas (f pl) esféricas y redondas
roulettes (f pl) et roues (f pl)
rotelle (f pl) e ruote

10 catalogue
Katalog (m)
catálogo (m)
catalogue (m)
catalogo (m)

11 catalyst
Katalysator (m)
catalizador (m)
catalyseur (m)
catalizzatore (m)

12 catalysts
Katalysatoren (m pl)
catalizadores (m pl)
catalyseurs (m pl)
catalizzatori (m pl)

13 caustic soda
Ätznatron (n), Natriumhydroxyd (n)
sosa (f) cáustica
soude (f) caustique
soda (f) caustica

14 cellophane
Zellophan (n)
celofán
cellophane (f)
cellofan (m)

15 cellulose
Zellulose (f), Cellulose (f), Zellstoff (m)
celulosa (f)
cellulose (f)
cellulosa (f)

16 cellulosic fibre
Zellulosefaser (f)
fibra (f) celulósica
fibre (f) cellulosique
fibra (f) cellulosica

17 centering
Zentrieren (n)
centraje (m)
centrage (m)
centratura (f)

18 centigramme
Zentigramm (n)
centígramo (m)
centigramme (m)
centigrammo (m)

1 centimetre
Zentimeter (m)
centímetro (m)
centimètre (m)
centimetro (m)

2 centrifugal
zentrifugal (adj)
centrífugo
centrifuge
centrifugo

3 centrifugal extractor
Zentrifugalextraktor (m)
extractor (m) centrífugo
extracteur (m) centrifuge
estrattore (m) centrifugo

4 centrifuge
Zentrifuge (f)
centrífugo (m)
centrifugeur (m)
centrifuga (f)

5 cerise
kirschfarben (adj)
color de cereza
rouge cerise
color celiego, fucsia

6 certificates of quality
Qualitätszertifikate (n pl)
certificados (m pl) de calidad
certificats (m pl) de qualité
certificati (m pl) di qualità

7 chain
Kette (f)
cadena (f), urdimbre (f)
chaîne (f)
catena (f)

8 chain drive
Kettenantrieb (m)
dirigido por cadena
commande (f) par chaîne
trasmissione (f) a catena

9 chalk
Kreide (f)
tiza (f)/yeso (m)
craie (f)
gesso (m)

10 chalks' fugitive
leichtentfernbare Kreidespuren (f pl)
tizas (f pl) fugitivas para tejidos
craies (f pl) fugaces [textiles]
elementi (m pl) di gesso non sostantivi

11 chamois [cloth]
Wildleder (n)
piel (f) de ante, gamuza (f)
peau (f) de daim
pelle (f) di daino

12 char
Verkohlen (n)
carbonizar (m)
carbonisé
carbonizzare

13 characteristic
Eigenschaft (f)
característica (f), propiedad (f)
caractéristique (f), propriété
caratteristica (f), proprietà (f)

14 charge
Aufladung (f), Ladung (f)
carga (f)
charge (f)
carica (f)

15 charge (v)
verladen, speisen, belasten,
beschweren
cargar, alimentar, aprestar
charger, alimenter
caricare

16 charges
Kosten (f pl), Spesen (f pl)
gastos (m pl), costes (m pl)
frais (m pl), charges (f pl)
spese (f pl)

17 charring
Karbonisation (f), Auskohlen (n)
carbonización (f)
carbonisage (m)
carbonizzazione (f)

18 chartreuse
chartreuse
chartreuse (m)
chartreuse (f)
verde pallido

1 check, control
Kontrolle (f)
control (m)
contrôle (m)
controllo

2 check, verification
Nachprüfung (f)
verificación (f)
vérification (f)
verifica (f)

3 check, square
Karo (n), Quadrat (n)
cuadro (m), cuadrado (m)
carreau (m), carré (m)
quadrato (m)

4 check pattern
Karomuster (n), Würfelmuster (n)
muestra (f) a cuadros
dessin (m) à petits carreaux
disegno (m) a quadretti

5 check-weighing
Kontrollwiegung (f)
pesada (f) de control
pesée (f) de contrôle
pesata (f) di controllo

6 cheese [bobbin]
Kreuzspule (f)
devanadera (f), bobina (f) cruzada
bobine (f) croisée
filo (m) avvolto su una bobina divisa,
rocca (f) crociata

7 cheese-cloth
Gaze (f)
tela (f) de devanadera
mousseline (f)
garza (f) commerciale

8 cheese-dyeing
Kreuzspulfärbung (f)
teñir (m) en devanadera
teinture (f) des bobines croisées
tintura (f) di rocche incrociate

9 cheese-winder
Kreuzspulmaschine (f)
bobinadora (f) de enrollamiento
cruzado
bobinoir (m) à bobines croisées
roccatrice (f) per rocche crociate

10 cheese-winding
Kreuzspulwickeln (n)
hilar (m) en devanadera
enroulement (m) des bobines croisées
incannatura (f) di rocche incrociate

11 chemical product
Chemikalie (f)
producto (m) quimico
produit (m) chimique
prodotto (m) chimico

12 chemical
chemisch (adj)
quimico
chimique
chimico

13 chemical analysis
chemische Analyse (f)
análisis (m) quimico
analyse (f) chimique
analisi (f) chimica

14 chemical formula
chemische Formel (f)
fórmula (f) quimica
formule (f) chimique
formula (f) chimica

15 chemical industry
Chemieindustrie (f)
industria (f) quimica
industrie (f) chimique
industria (f) chimica

16 chemical product
Chemikalie (f)
producto (m) quimico
produit (m) chimique
prodotto (m) chimico

17 chemical property
chemische Eigenschaft (f)
propiedad quimica
qualité (f) chimique
proprietà (f) chimica

18 chemical reaction
chemische Reaktion (f)
reacción (f) quimica
réaction (f) chimique
reazione (f) chimica

1 chemicals – textiles
Chemikalien (f pl) – für Textilien (f pl), kunstoffe (m pl)
productos quімicos para la industria textil
produits (m pl) chimiques [textiles]
sostanze (f pl) chimiche per il settore tessile

2 chemistry
Chemie (f)
quimica (f)
chimie (f)
chimica (f)

3 chenille [cloth]
Chemille (f)
chenilla (f), felpilla (f)
chenille (f)
ciniglia (f)

4 cheque
Scheck (m)
cheque (m)
chèque (m)
assegno (m)

5 Cheviot wool
Cheviotwolle (f)
lana (f) cheviot
cheviote (f)
lana (f) cheviot

6 Cheviot weave
Cheviotgewebe (n)
tejido (m) cheviot
armure (f) cheviote
tessuto (m) cheviot

7 chiffon [cloth]
Chiffon (m)
chiffón (m)
chiffon (m)
chiffon (m)

8 China grass [fibre]
Chinagras (n), Ramie (f)
ramio (m)
ramie (f)
ramié (m)

9 chinchilla [cloth]
Chinchilla (n)
chinchillá (f)
chinchilla (m)
cincillà (m)

10 chintz [cloth]
Chintz (m)
chintz (m), zaraza (f), glaseda (f)
chintz
cinz (m)

11 chlorate
Chlorat (n)
clorato (m)
chlorate (m)
clorato (m)

12 chloride
Chlorid (n)
cloruro (m)
chlorure (m)
cloruro (m)

13 chlorination
Chlorung (f), Chloren (n), Chlorierung (f)
cloración (f), clorización, clorado (m)
chloration (f), chlorage (m)
clorazione (f)

14 chlorine
Chlor (n)
cloro (m)
chlore (m)
cloro (m)

15 chlorofibres
Chlorfasern (f pl)
fibras (f pl) de cloro
chlorofibres (f pl)
clorofibre (f pl)

16 chromatic
chromatisch (adj)
cromático
chromatique
cromatico

17 chrome
Chrom (n)
cromo (m)
chrome (m)
cromo (m)

18 chrome-dyeing
Chromfarben (n)
teñir (m) con cromo
teinture (f) au chrome
tintura (f) al cromo

1 chromophore
Chromophor (m)
cromófero (m)
chromophore (m)
cromoforo (m)

2 cinnamon
zimtbraun (adj)
cáñamo (m)
cannelle (f) de Padang
cannella (f) [colore]

3 circle
Kreis (m)
círculo (m)
cercle (m)
cerchio (m)

4 circular
kreisförmig, rund (adj)
circular
circulaire
circolare

5 circular knitting
Rundstricken (n)
hacer (m) puntos en forma circular
tricotage (m) circulaire
lavorazione (f) tubolare a maglia circolare

6 circular knitting machine
Rundwirkmaschine (f), Rundstrickmaschine (f)
tricotosa (f) circular, máquina (f) circular
tricoteuse (f) circulaire, machine (f) circulaire pour tricoter
macchina (f) circolare, macchina (f) circolare per maglieria

7 circular looping machine
Rundkettelmaschine (f)
máquina (f) circular de remallar, máquina (f) circular de remallado
remailleuse (f) circulaire
rimagliatrice (f) circolare

8 circular purl knitting machine
Links-Links-Rundstrickmaschine (f)
máquina (f) circular links-links
métier (m) circulaire à mailles retournées
macchina (f) circolare per maglia rovesciata

9 circular rib knitting machine
Rechts-Rechts-Rundstrickmaschine (f)
máquina (f) circular de punto acanalado
Métier (m) circulaire à côtes
macchina (f) tubolare a coste

10 circulate (v)
umlaufen, zirkulieren
circular
circuler
circolare

11 circulation
Umlauf (m), Kreislauf (m), Zirkulation (f)
circulación (f)
circulation (f)
circolazione (f)

12 circumference
Kreisumfang (m)
circunferencia (f)
circonférence (f)
circonferenza (f)

13 claim
Anfechtung (f), Reklamation (f)
reclamación (f)
contestation (f)
contestazione (f)

14 clarification
Klärung (f)
clarificación (f)
clarification (f)
chiarificazione (f)

15 classification
Klassierung (f), Sortierung (f)
clasificación (f)
classement (m)
classificazione (f)

16 classify (v)
einteilen, klassieren, sortieren
clasificar, graduar
classifier, classer
classificare

17 classing
Klassierung (f), Sortierung (f)
clasificación (f)
classement (m)
classificazione (f)

1 clean
sauber, rein
limpio
propre
pulito

2 cleaning cloth
Putztuch (n), Wischtuch (n)
trapo (m) de limpieza
chiffon (m) de nettoyage
strofinaccio (m)

3 clearance
Verzollung (f)
despacho (m) de aduana
dédouanement (m)
sdoganamento (m)

4 clearing, purification
Klären, Läuterung (f), Reinigung (f)
purificación (f)
purification (f)
depurazione (f)

5 clippings
Stoffschnitzel (m pl), Stoffabfälle (m pl)
recortes (m), retales (m pl)
laine (f) de peddler, déchets (m pl)
ritagli (m pl)

6 close
nah, anliegend (adj)
vecino, cercano, contiguo
voisin, près
vicino

7 closed shed [loom]
geschlossenes Fach (n),
Geschlossenfach (n)
taller (m) [de telares], calada (f)
cerrada
foule (f) fermée
angolo (m) di inserzione chiuso, passo
(m) chiuso

8 closely woven fabric
dichte Ware (f)
tejido (m) compacto, tejido (m) denso
tricot (m) serré
tessuto (m) compatto

9 cloth
Tuch (n), Stoff (m)
tela (f)
tissu (m), étoffe (f)
tessuto (m)

10 cloth – abba
Abbatuch (n)
tela (f) de urdimbre
tissu (m) abba
tessuto (m) – di bassa qualità

11 cloth – acrylic
Acryltuch (n)
tela (f) – acrílica
tissu (m) acrylique
tessuto (m) – acrilico

12 cloth – afgalaine
Afghalaine (m)
tela (f) – afgalaine
tissu (m) afgalaine
tessuto (m) – afgalaine

13 cloth – agents
Stoff (m) – Mittel (n pl)
agentes (m pl) de telas
tissu (m) – agents (m pl)
tessuto (m) – agenti (m pl)

14 cloth – alpaca
Alpakastoff (m)
tela (f) de alpaca
tissu (m) alpaga
tessuto (m) – alpaca

15 cloth – apron
Schürzentuch (n)
telas (f pl) para delantales
étoffe (f) pour tablier
tessuto (m) – nastro trasportatore
a piastre

16 cloth – asbestos
Gewebe (n) – Asbest (n)
tela (f) de amianto
tissu (m) d'amiante
tessuto (m) – amianto

17 cloth – backed
kaschiertes Tuch (n)
telas (f pl) forradas
tissu (m) renforcé
tessuto (m) – rinforzato

18 cloth – Barathea
Baratheastoff (m)
tela (f) Baratea
tissu (m) barathéa
tessuto (m) – incrociato con disegni
in rilievo

1 cloth – Bedford cord
Bedfordkordstoff (m)
tela (f) tipo pana bedford
tissu (m) bedford
tessuto (m) – Bedford, Tessuto diagnale

2 Cloth – billiard and pool
Billiard – und Pooltuch (n)
tela (f) para mesas de billar y trucos
tapis (m) de billard
panno (m) – per bigliardo e giochi simili

3 cloth – blanket
Deckenstoff (m)
tela (f) para mantas
tissu (m) pour couverture
tessuto (m) – coperta

4 cloth – blankets – polyester cellular
Gewebe (n) – Decken (f pl) – zelluläres Polyester (n)
tela (f) de poliéster celular para mantas
tissu (m) – couvertures en polyester cellulaire
tessuto (m) – per coperte – poliestere cellulare

5 cloth – blankets – woollen (m)
Gewebe (n) – Decken (f pl) – aus Wolle (f)
tela (f) de lana para mantas
tissu (m) – couvertures en laine
tessuto (m) – per coperte – di lana

6 cloth – blankets – wool mixture
Gewebe (n) – Decken (f pl) – Mischwolle (f)
tela (f) de mezclas de lana para mantas
tissu (m) – couvertures en laine mélangée
tessuto (m) – per coperte – misto lana

7 cloth – blazer
Blazertuch (n)
tela (f) para chaquetas ligeras
flanelle (f)
tessuto (m) – per giacche sportive

8 cloth – blowing
Stoff (m) – Dampfblasen (n)
tela (f) prensada con vapor
soufflage (m) des poils, ébouriffage (m)
tessuto (m) – soffiaggio

9 cloth – bouclé
Boucléstoff (m)
tela (f) boucle
tissu (m) bouclé
tessuto (m) – boucle

10 cloth – brushing
Stoff (m) – Bürsten (n)
tela (f) cepillada
brossage (m) de tissu
tessuto (m) – spazzolatura

11 cloth – bunting
Fahnentuch (n)
lanilla (f)
molleton (m) à drapeaux
tessuto (m) – stamigna

12 cloth - calico
Kattunstoff (m)
tela (f) de calicó
cotonnade (f)
tessuto (m) – calico

13 cloth – camel hair
Kamelhaarstoff (m)
tela (f) de pelo de camello
poil (m) de chameau
tessuto (m) – pelo di cammell

14 cloth – camel hair/wool
Kamelhaarwollstoff (m)
tela (f) de pelo de camello y lana
poil (m) de chameau/laine
tessuto (m) – pelo di cammello/lana

15 cloth – canvas tapestry and rug
Gewebe (n) – Leinenwandteppiche und kleine Teppiche [Vorlage]
tela (f) de lona para tapicería y alfombrillas
tissu (m) – toile de tapisserie et moquettes
tessuto (m) – per canovacci, tappezzeria e tappetini

16 cloth – cap and hat
Mützen – und Hutstoff (m)
tela (f) para gorras y sombreros
étoffe (f) pour casquettes et chapeaux
tessuto (m) – cappelli e berretti

17 cloth – Cashmere
Kaschmirstoff (m)
tela (f) de casimir
cachemire (m)
tessuto (m) – Cashmere

1 cloth – Cashmere/silk
Halbseidenkaschmirstoff (m)
tela (f) de casimir y seda
cachemire (m) soie (f)
tessuto (m) – Cashmere/seta

2 cloth – Cashmere/wool
Wollkaschmirstoff (m)
tela (f) de casimir y lana
cachemire (m) laine (f)
tessuto (m) – Cashmere/lana

3 cloth – cavalry twill – blended
Kavallerietwillstoff (m) gemischt
tela (f) de caballería cruzada y mezclada
croisé lourd – mélangé
tessuto (m) – diagonale – misto

4 cloth – cavalry twill – wool
Kavallerietwill-Wollstoff (m)
tela (f) de caballería cruzada con lana
croisé lourd – laine
tessuto (m) – diagonale – lana

5 cloth – centrifuging
Stoffabschleudern (n)
tela (f) centrifugada
centrifugation (f) [de tissu]
tessuto (m) – centrifugazione

6 cloth – Cheviot
Cheviotstoff (m)
tela (f) cheviot
cheviote (f)
tessuto (m) – panno Cheviot

7 cloth – cleaning
Stoff (m) – Reinigen (n)
paños (m pl) para limpiar
toile (f) nettoyeuse
tessuto (m) – pulitura

8 cloth – clerical vestments
sakrale Bekleidung (f), Priester-bekleidung (f)
telas (f pl) para vestidos sacerdotales
vêtements (m pl) cléricaux
tessuto (m) – paramenti liturgici (abiti sacri)

9 cloth – cockling
Kräuselungstoff (m)
tela (f) para arrugarse
crispage (m) de tissu
tessuto (m) – arricciatura

10 cloth – coffin linings
Sargtuch (n)
tela (f pl) para forrar féretros
revêtement (m) intérieur de cercueil
tessuto (m) – foderatura, rivestimento feretri

11 cloth – collar Melton
Stoff (m) – kragenmelton (m)
telas (f pl) para los cuellos tipo melton
col (m) melton
tessuto (m) – Melton per colletti

12 cloth construction
Gewebestruktur (f)
textura (f) del tejido
structure (f) du tissu
struttura (f) del tessuto

13 cloth – contract
Tuchvertrag (m)
telas (f pl) de contrato
tissu (m) par contrat
tessuto (m) – contratto

14 cloth – corduroys
Kordsamtstoff (m)
tela (f) de pana
velours (m) côtelé
tessuto (m) – velluto a coste

15 cloth – corset
Miederstoff (m)
tela (f) para fajas
tissus (m pl) pour corsets
tessuto (m) – per corsetti

16 cloth – costume
Trachtenstoff (m)
tela (f) para trajes
tissu (m) pour costumes
tessuto (m) – per costumi

17 cloth – cotton
Baumwollstoff (m)
tela (f) de algodón
étoffe (f) de coton
tessuto (m) – cotone

18 cloth – cotton loomstate
Baumwollrohware (m)
tela (f) de algodón directa del telar
coton (m) écru
tessuto (m) – gieggo da telaino

1 cloth – cotton repp 100%
Stoff (m) – Baumwollrips 100%
tela (f) de algodón rematada en ambos lados 100%
reps (m) en coton à 100%
tessuto (m) – 100% cotone a coste repp

2 cloth – cotton, undyed
Stoff (m) – naturfarbene Baumwolle (f)
tela (f) de algodón no teñido
étoffe (f) de coton non-teinte
tessuto (m) – cotone non tinto, cotone grezzo

3 cloth – covert
Stoff (m) – Covercoat (m)
tela (f) asargada
tissu (m) cover
tessuto (m) – specie di tweed

4 cloth – crepe
Kreppstoff (m)
tela (f) de crepé
tissu (m) crêpe
tessuto (m) – crespo

5 cloth – creams/cricket and tennis
Stoff (m) – Sportwollwaren (f pl)
tela (f) blanca amarillenta para ropa de críquet y de tenis
tissu (m) blanc crème pour cricket et tennis
tessuto (m) – color panna – per cricket e tennis

6 cloth – cropping
Stoffscheren (n)
tela (f) recortada
tondage (m) de tissu
tessuto (m) – cimatura

7 cloth – cut pile
Sotff(m)–geschnittenes Florgewebe (n), Schnittflorgewebe (n)
tela (f) de pelusa cortada
tissu (m) à poil coupé
tessuto (m) – pelo corto

8 cloth – damping
Stoff(m)–Befeuchten(n), stoffdämpfen (n)
tela (f) rehumedecida
décatissage (m) [de tissu]
tessuto (m) – inumidimento

9 cloth – deck chair
Liegestuhlstoff (m)
tela (f) para sillas plegables
tissu (m) pour transat
tessuto (m) – per sedie a sdraio

10 cloth density
Warendichte (f), Stoffdichte (f), Gewebedichte (f)
densidad (f) del tejido
densité (f) du tissu
densità (f) del tessuto, fittezza (f) del tessuto

11 cloth – dish [dishcloth]
Geschirrtuch (n)
paños (m pl) para fregar platos
tissu (m) pour lavettes
tessuto (m) – strofinaccio, canovaccio per piatti

12 cloth – doeskins
Stoff (m) – Döskin (m), Doeskin (m) Doeskinstoff (m)
tela (f) de piel de cierva
doeskin (m)
tessuto (m) – pelli di daina

13 cloth – double
Stoff (m) – Doppelgewebe (n)
tela (f) doble
étoffe (f) matelassée
tessuto (m) – doppio

14 cloth – double plains
Stoff (m) – leinwandbindiges Doppelgewebe (n)
tela (f) rematada en ambos lados
tissu (m) double-lisse
tessuto (m) – doppio non lavorato

15 cloth – drawn finish
Stoff (m) – gefinishter Stoff (m)
tela (f) levadiza
apprêt (m) étiré
tessuto (m) – rifinitura, stirato

16 cloth - dry finishing
Stoff (m) – Trockenappretur (f)
tela (f) rematada en seco
apprêt (m) sec
tessuto (m) – rifinitura a secco

17 cloth – duck
Duckstoff (m) zeltstoff (m)
tela (f) de dril
toile (f) lourde
tessuto (m) – tela olona

1 cloth – duffle
Düffelstoff (m)
paño (m) de lana basta
molleton (m)
tessuto (m) – di lana grezza

2 cloth edge
Warenrand (m)
borde (m) del tejido
bord (m) du tissu
bordo (m) del tessuto

3 cloth – examination
Stoff (m) – Untersuchung (f)
tela (f) controlada
vérification (f) [de tissu]
tessuto (m) – esame

4 cloth examing
Warenschau (f), Gewebeprüfung (f), Repassieren (n)
examen (m) de las telas, análisis (m) de los tejidos, repaso (m) de control de tejidos, verificación (f) de las telas
vérification (f) des pièces, examen (m) de tissus, repassage (m) de contrôle de tissus
ispezione (f) delle pezze, esame (m) dei tessuti, ripassatura (f)

5 cloth – felt
Gewebe (n) – Filz (m), Filtegewebe (n)
tela (f) de fieltro
tissé (m) feutre
tessuto (m) – feltro

6 cloth – filters
Filterstoff (m)
paño (m) para filtros
tissu (m) pour filtres
tessuto (m) – per filtri

7 cloth – filters anti-static
Gewebe (n) für Antistatikfilter
filtros (m pl) anti-estáticos de tela
filtres (m pl) en tissu antistatique
tessuto (m) – per filtri antistatici

8 cloth – fire-resistant
feuerbeständiges Tuch (n)
tela (f) resistente al fuego
tissu (m) ininflammable
tessuto (m) – resistente al fuoco

9 cloth – flag
Fahnentuch (n)
tela (f) para banderas
tissu (m) pour drapeaux
tessuto (m) – per bandiere

10 cloth – flameproof
feuerfester Stoff (m)
tela (f) ininflamable
tissu (m) ignifuge
tessuto (m) – ininflammabile

11 cloth – flannel-woollen
Stoff (m) – Flannel (m)
tela (f) de franela y lana
flanelle (f) de laine
tessuto (m) – flanella – lana

12 cloth – flannel-worsted
Stoff (m) – Kammgarnflannel (m)
tela (f) de franela y estambre
flanelle (f) peignée
tessuto (m) – flanella – pettinato

13 cloth – flax/linen
Leinenstoff (m)
tela (f) de lino/lienza
lin (m)
tessuto (m) – fibra di lino/lino

14 cloth – folding
Stoff (m) – Falten (n)
tela (f) plegable
assemblage (m) parallèle, pliage (m)
piegatura del tessuto

15 cloth – footwear
Fußbekleidungsstoffe (m pl)
tela (f) para calzados
articles (m pl) chaussants
tessuto (m) – per calzature

16 cloth – furnishing
Wohntextilienstoffe (m pl)
tela (f) para muebles
tissu (m) d'ameublement
tessuto (m) – per arredamento

17 cloth – gaberdine
Gabardinestoff (m)
tela (f) para gabardinas
tissu (m) gabardine
tessuto (m) – gabardine

1 cloth – glass
Gewebe (n) – Glas (n), Glasgewebe (n)
tela (f) de vidrio
tissu (m) de verre textile
tessuto (m) – vetro

2 cloth guide
Stoffbahnführer (m), Gewebeführer (m), Warenbahnführer (m)
guíatelas (m)
conducteur (m) de nappe de tissu, guide-tissu (m)
guida-pezza (m), guida-tessuto (m)

3 cloth – hair
Gewebe (n) – Haar (n), Haargewebe
tela (f) de pelo
étoffe (f) de crin
tessuto (m) di pelo

4 cloth – Harris tweed
Stoff (m) – Harris Tweed (m)
paño (m) de harris
tweed (m) Harris
tessuto (m) – Harris tweed

5 cloth – headlining
Autodachfütterungsstoff (m) Autodachfutterstoff (m)
tela (f) para forros tiesos
revêtement (m) du toit
tessuto (m) – da rivestimento

6 cloth – hopsack
loser Wollstoff (m), Hopsack
tejido (m) de cesta
cheviote (f) grossière
tessuto (m) – tela grezza

7 cloth – indigo [colour]
Stoff (m) – Indigo
tela (f) de añil
tissu (m) indigo
tessuto (m) – color indaco

8 cloth – industrial
Industriestoff (m)
tela (f) industrial
étoffe (f) industrielle
tessuto (m) – industriale

9 clothing
Kleidung (f), Bekleidung (f)
indumentaria (f), ropa (f) de vestir, vestimenta (f)
habillement (m), vêtement (m pl)
abbigliamento (m), vestiario (m)

10 clothing, trimming, packing
Besatz (m), Ausstaffierung (f), Belag (m), Dichtung (f)
adorno (m), guarnición (f), empaquetadura (f)
garniture (f), joint (m)
guarnizione (f)

11 cloth – interlining
Stoff (m) – Zwischenfutter (n)
tela (f) para forros tiesos e internos
doublure (f) intermédiaire
tessuto (m) – interfodera

12 cloth – jacketings
Stoff (m) für Jacken, Jackets (f pl)
tela (f) para chaquetas
tissu (m) pour vestes
tessuto (m) – per giacche

13 cloth – jacketings/camel hair
Stoff (m) für Kamelhaarjaken (f pl)
tela (f) de pelo de camello para chaquetas
poil (m) de chameau pour vestes
tessuto (m) – per giacche – pelo di cammell

14 cloth – jacketings/cashmere
Stoff (m) für Kashmirjaken (f pl)
tela (f) de cachemir para chaquetas
cachemire (m) pour vestes
tessuto (m) – per giacche – cashmere

15 cloth – jacketings/mixture
Stoff (m) für Jacken aus Mischgewebe
tela (f) mezclada para chaquetas
tissu (m) mixte pour vestes
tessuto (m) – per giacche – misto

16 cloth – jacketings/natural fibres
Stoff (m) – Naturfaserummantelung aus naturfasern für Jacken (f pl)
tela (f) de fibras naturales para chaquetas
tissu (m) pour vestes en fibres naturelles
tessuto (m) – per giacche – fibre naturali

17 cloth – jacketings/pure wool
Stoff (m) aus reiner Wolle für Jacken (f pl)
tela (f) de lana pura para chaquetas
tissu (m) pour vestes en pure laine
tessuto (m) – per giacche – pura lana

1 cloth – jacketings/woolmark
Stoff (m) für Jacken aus Wollseigelqualität
tela (f) de lana virgen para chaquetas
tissu (m) pour vestes – woolmark
tessuto (m) – per giacche – marchio di pura lana

2 cloth – jacketings/wool/cashmere blends etc.
Stoff (m) – Woll/Kaschmirmischummantelung (f) aus Wolle Kashmirstoff für jacken
tela (f) de mezclas de lana/casimir etc. para chaquetas
tissu (m) pour vestes en laine/cachemire, mélanges etc.
tessuto (m) – per giacche – misto lana/cashmere ecc.

3 cloth – Jacquard
Stoff (m) – Jacquard (m)
tela (f) jacquard
tissu (m) jacquard
tessuto (m) – jacquard

4 cloth – knitted
gewirkter Stoff (m)
tela (f) de puntos
tricotages (m pl)
tessuto (m) - per fodera

5 cloth – ladies' coatings
Damenmantelstoff (m)
tela (f) para abrigos de senoras
tissu (m) pour manteaux pour femmes
tessuto (m) – soprabiti da donna

6 cloth – lambswool
Lammwollestoff (m)
tela (f) de lana de cordero
laine (f) d'agneau
tessuto (m) – lambswool

7 cloth – laundry machine
Wäschereinmaschinenstoff (m)
tela (f) para máquinas de lavanderías
étoffe (f) pour machines de blanchisserie
tessuto (m) – per macchine da lavanderia

8 cloth – leisurewear
Stoff (m) für Freizeitbekleidung (f)
tela (f) para ropa informal
tissu (m) pour vêtements de loisir
tessuto (m) – abbigliamento per il tempo libero

9 cloth – linen
Leinenstoff (m), Leinentuch (n)
lienzo (m)
étoffe (f) lin
tessuto (m) – lino

10 cloth – linen/apparel
Leinenbekleidungsstoff (m)
género (m) de lino para ropa
lin (m) pour vêtements
tessuto (m) – lino – per abiti

11 cloth – linen/damask
Leinendamaststoff (m)
lienzo (m) de damasco
lin (m) damas
tessuto (m) – lino – damasco

12 cloth – linen household
Hausewäschestoff (n)
ropa (f) blanca de casa
linge (m) de maison
tessuto (m) – lino – uso domestico

13 cloth – lining
Futterstoff (m)
tejido (m) para forros
doublure (f) en drap
tessuto (m) – per fodera

14 cloth measuring and folding machine
Meß und Legemaschine (f) für Gewebe
máquina (f) medidora-plegadora de tejidos
plieuse-métreuse (f) mécanique pour tissus
macchina (f) misuratrice e piegatrice della pezza

15 cloth – melange
Melangestoff (m)
tela (f) mezclada
étoffe-mélange (f)
tessuto (m) – melange

16 cloth – Melton
Stoff (m) – Melton (m)
tela (f) melton
tissu (m) melton
tessuto (m) – Melton

17 cloth – mending
Stoff (m) – Reparieren (n)
tela (f) remendada
tissu (m) raccommodage
tessuto (m) – rammendo

1 cloth – mens' coatings
Herrenmantelstoff (m)
tela (f) para abrigos de caballeros
tissu (m) pour pardessus
tessuto (m) – soprabiti da uomo

2 cloth – merchant
Stoffhändler (m)
comerciante (m) de tela
négociant (m) en draps
tessuto (m) – commerciante

3 cloth – merchants
Stoffhändler (m pl)
comerciantes (m pl) de tela
négociants (m pl) en draps
tessuto (m) commercianti

4 cloth – merchant convertors
Stoffhändler Konverter (m pl), Manipulanten (m pl)
telas - convertidores
négociants (m pl) – transformation
tessuto (m) – commerciali, commercianti, convertitori

5 cloth – mohair
Mohairstoff (m)
tela (f) de angora
mohair (m)
tessuto (m) – mohair

6 cloth – mohair/blends
Mohairmischstoffe (m pl)
tela (f) mezclada de angora
mélanges (m pl) mohair
tessuto (m) – mohair/misto

7 cloth – mouflon
Stoff (m) – Moufflons (m pl)
paño (m) de musmón
tissu (m) mouflons
tessuto (m) – mufflani

8 cloth – neckties
Krawattenstoff (m)
tela (f) para corbatas
tissu (m) pour cravates
tessuto (m) – cravatte

9 cloth – non-woven
Faservliesstoff (m)
tela (f) no tejida
étoffe (f) nappée
tessuto (m) – non tessuto

10 cloth – overcoating, woollen
wollener Mantelstoff (m)
tela (f) de lana para abrigos
tissu (m) pour pardessus en laine
tessuto (m) – lana da soprabiti

11 cloth – overcoating, worsted
Kammgarnmantelstoff (m)
tela (f) de estambre para abrigos
tissu (m) pour pardessus en laine peignée
tessuto (m) – pettinato da soprabiti

12 cloth – panama
Panamastoffe (m pl)
tela (f) para jipijapas
tissu (m) pour panamas
tessuto (m) – panama

13 cloth – pile
Florstoff (m)
tela (f) de pelusa
tissu (m) velouté
tessuto (m) – pelo

14 cloth – polyester
Stoff (m) – Polyester (m)
tela (f) de poliéster
tissu (m) polyester
tessuto (m) – poliestere

15 cloth – polyester/cotton
Stoff (m) – Polyester/Baumwolle (f)
tela (f) de poliéster/algodon
tissu (m) polyester/coton
tessuto (m) – poliestere/cotone

16 cloth – polyester/viscose
Stoff(m) – Polyester/Viskose (f)
tela (f) de poliéster/viscosa
tissu (m) polyester/viscose
tessuto (m) – poliestere/viscosa

17 cloth – polypropylene
Stoff (m) – Polypropylene (n)
tela (f) de polipropileno
tissu (m) polypropylène (m)
tessuto (m) – polipropilene

18 cloth – pressing
Stoffbügeln (n)
tela (f) prensada
tissu (m) pressage
tessuto (m) – stiratura

1 cloth – printed
bedruckter Stoff (m)
tela (f) estampada
tissu (m) imprimé
tessuto (m) – stampato

2 cloth – raised
gerauhter Stoff (m)
paño (m) perchado
tissu (m) gratté
tessuto (m) – garzato

3 cloth – raised and sueded
gerauhter und velourierter Stoff (m)
paño (m) perchado con aspecto de gamuza
tissu (m) gratté et suedine
tessuto (m) – garzato e scamosciato

4 cloth – rayon
Stoff (m) – Reyon (m)
tela (f) de rayón
tissu – rayonne (f)
tessuto raion (m)

5 cloth – rayon/cotton brocade
Stoff (m) – Reyon/Baumwollbrokat (m)
tela (f) de rayón/brocada de algodón
tissu-rayonne (f)/brocart (m)
tessuto (m) – broccato rayon/cotone

6 cloth – rep worsted
Stoff (m) – Kammgarnrips (m)
tela (f) de estambre rematada en ambos lados
reps (m) en laine peignée
tessuto (m) – a repp pettinato

7 cloth – reversible
doppelseitiger Stoff (m)
tela (f) reversible
tissu (m) réversible
tessuto (m) – double face

8 cloth – roll
Stoffrolle (f), Warenwicker (m)
rollo (m) de tela, rollo (m) de tejido
rouleau (m) de tissu
tessuto (m) – rotolo

9 cloth – rolling
Stoff (m) – Walzen (n), Walzentuch (n)
tela (f) arrollada
enroulement (m) de tissu
tessuto (m) – arrotolamento (m)

10 cloth – Russell cord
Stoff (m) – Russellkord (m)
tela (f) de pana russell
reps (f) de laine
tessuto (m) – a corte Russell

11 cloth – satin
Gewebe (n) – Satin (n), Satingewebe
tela (f) de raso
satin (m)
tessuto (m) – satin

12 cloth – satin worsted
Stoff (m) – Kammgarnsatin (m)
tela (f) de raso/estambre
satins (m pl) laine peignée
tessuto (m) – satin pettinati

13 cloth – stole
Gewebe (n) für Stolen (f pl)
tela (f) para estolas
tissu (m) pour étoles
tessuto (m) – per stole

14 cloth – Saxony
Stoff (m) aus Elektorawolle (f)
tela (f) para saxony [tejido fino]
tissu (m) – saxony
tessuto (m) – di lana di Sassonia, saxony

15 cloth – scarf
Halstuchstoff (m)
tela (f) para bufandas
tissu (m) pour écharpes
tessuto (m) – sciarpa

16 cloth – scarf/pure wool
Schalwollstoff (m)
tela (f) de lana pura para bufandes
tissu (m) pour écharpes – pure laine
tessuto (m) – di pura lana per sciarpa

17 cloth – scouring
Stoff (m) – Reinigen (n)
tela (f) descrudado
lavage (m) de tissu à fond
tessuto (m) – purga

18 cloth – serges
Wollsergen (f pl)
tela (f) de estameña
tissu (m) – serges
tessuto (m) – saie

1 cloth – shawl
Schalsstoff (m)
tela (f) para chales
tissu (m) pour châles
tessuto (m) – scialli

2 cloth – sheeting/linen
Bettuchleinen (n)
lienzo (m) para sábanas
tissu (m) brut de coton
tessuto (m) – per lenzuola – biancheria

3 cloth – Shetland
Stoff(m) – Shetland (m)
tela (f) shetland
tissu (m) de laine Shetland
tessuto (m) – Shetland

4 cloth – silk
Seidenstoff (m)
tela (f) de seda
tissu (m) de soie
tessuto (m) – seta

5 cloth – silk/wool
Wollseidestoff (m)
tela (f) de seda/lana
tissu (m) de soie/laine
tessuto (m) – seta/lana

6 cloth – skirt
Rockstoff (m)
tela (f) para faldas
tissu (m) pour jupes
tessuto (m) – per gonna

7 cloth – skirt/pure wool
reiner Wollrockstoff (m)
tela (f) de lana pura para faldas
tissu (m) pour jupes – pure laine
tessuto (m) per gonna – pura lana

8 cloth – skirt/woolmark
Wollsiegelrockstoff (m)
tela (f) de lana virgen para faldas
tissu (m) pour jupes – woolmark
tessuto (m) per gonna – marchio pura lana

9 cloth – sports suiting
Sportanzugstoff (m)
tela (f) para trajes deportivos
tissu (m) pour vêtements de sport
tessuto (m) – per abiti sportivi

10 cloth – suiting/linen
Leinenanzugstoff (m)
tela (f) para trajes de lienzo
lin (m) de confection
tessuto (m) – per abiti – lino

11 cloth – suiting/woollen
Wollanzugstoff (m)
tela (f) para trajes de lana
étoffe (f) de laine de confection
tessuto (m) – per abiti – lana

12 cloth – suiting/worsted
Kammgarnanzugstoff (m)
tela (f) para trajes de estambre
étoffe (f) de laine peignée de confection
tessuto (m) – per abiti – pettinato

13 cloth – tartan mixture
Schottenkar-Mischungstoff (m)
tela (f) de tartán mezclado
tissu (m) mixte écossais
tessuto (m) – misto tartan

14 cloth – tartan/pure wool
Schotenkaro (n) reiner-Wollstoff (m)
tela (f) de tartán de lana pura
tissu (m) écossais pure laine
tessuto (m) – tartan – pura lana

15 cloth – tennis ball
Tennisballstoff (m)
tela (f) para pelotas de tenis
tissu (m) pour balle de tennis
tessuto (m) – per palle da tennis

16 cloth – terylene/wool
Terylen/Wollstoff (m)
tela (f) de tergal/lana
tissu (m) de terylène/laine
tessuto (m) – terilene/lana

17 cloth – terylene/worsted
Terylen/Kammgarnstoff (m)
tela (f) de tergal/estambre
tissu (m) de terylène/laine peignée
tessuto (m) – terilene/pettinato

18 cloth testing
Gewebeprüfung (f)
análisis (m) de los tejidos, examen (m) de las telas
examen (m) de tissus
esame (m) dei tessuti

1 cloth – textured polyester
texturierter Polyesterstoff (m)
tela (f) de poliéster tejido
tissu (m) de polyester texturé
tessuto (m) – poliestere tessuto

2 cloth – ticking
Bettkattunstoff (m)
tela (f) de terliz
coutil (m) pour literie
tessuto (m) – tralicci

3 cloth – tie-lining
Krawattenfutterstoff (m)
tela (f) para forrar corbatas
doublures (f pl) de cravates
tessuto (m) – fodere per cravatte

4 cloth – topcoatings
Mantelstoffe (m pl)
capas (f pl) impermeables para tela
tissu (m) pour paletot
tessuto (m) per soprabiti

5 cloth – topcoatings/camel hair
Kamelhaarmantelstoffe (m pl)
capas (f pl) impermeables para tela de pelo de camello
poil (m) de chameau pour paletot
tessuto (m) per soprabiti – pelo di camello

6 cloth – topcoatings/cashmere
Kaschmirmantelstoffe (m pl)
capas (f pl) impermeables para tela de cachemir
cachemire (m) pour paletot
tessuto (m) per soprabiti – cashmere

7 cloth – topcoatings/mixture
Mischgewebemantelstoff (m)
capas (f pl) impermeables para tela mezclada
tissu (m) mixte pour paletot
tessuto (m) per soprabiti – misto

8 cloth – topcoatings/natural fibres
Naturfasermantelstoff (m)
capas (f pl) impermeables para tela de fibras naturales
tissu (m) de fibres naturelles pour paletot
tessuto (m) per soprabiti – fibre naturali

9 cloth – topcoatings/wool/cashmere blends etc.
Wolle/Kaschmir – Mischgewebemantel-stoff (m)
capas (f pl) impermeables para tela de mezclas de lana/cachemir etc.
étoffe (f) de laine/cachemire/mélanges etc. pour paletot
tessuto (m) soprabiti – lana/misto cashmere

10 cloth – topcoatings/woolmark
Wollesiegelmantelstoff (m)
capas (f pl) impermeables de tela de pura lana virgen
étoffe (f) woolmark pour paletot
tessuto (m) per soprabiti – marchio di pura lana

11 cloth – towelling
Handtuchstoff (m)
tela (f) para toallas
tissu (m) éponge
tessuto (m) – per asciugamani

12 cloth – toys
Stoff (m) für spielzeuge
tela (f) para juguetes
tissu (m) pour jouets
rela tessuto (m) – per giocattoli

13 cloth – travel rugs
Gewebe (n) – Reisedecken (f pl)
tela (f) para mantas de viaje
tissu (m) pour plaids
tessuto (m) – tappetini da viaggio

14 cloth – trevira/worsted/mohair
Trevira/Kammgarn/Mohairstoff (m)
tela (f) de trevira/estambre/mohair
étoffe (f) de laine peignée/mohair/ trevire
tessuto (m) – trevira/pettinato/mohair

15 cloth – trevira/worsteds
Trevira/Kammgarnstoffe (m pl)
tela (f) de trevira/estambre
étoffe (f) de laine peignée/trevire
tessuto (m) – trevira/pettinato

16 cloth – tropical
Tropenstoff (m)
tela (f) ligera
tissu (m) tropicalisé
tessuto (m) –leggero

1 cloth – trousering
Hosenstoff (m)
tela (f) para pantalones
tissu (m) pour pantalon
tessuto (m) – per calzoni

2 cloth – tweeds
Tweedstoffe (m pl)
paño (m) de lana de dos colores
tissu-tweeds (m)
tessuto (m) – tweed (spigato)

3 cloth – twist
Zwirnstoff (m) Zwirntuch (n)
tela (f) de trenzas
retors (m) de tissu
tessuto (m) – rotorto

4 cloth – undyed
naturfarbene Stoffe (m pl)
tela (f) no teñida
tissu (m) non-teint
tessuto (m) – non tinto

5 cloth – uniform
Uniformstoff (m)
tela (f) para uniformes
tissu (m) pour uniformes
tessuto (m) – per uniformi

6 cloth – upholstery
Möbelstoff (m)
tela (f) para tapicerías
tissu (m) d'ameublement
tessuto (m) – per arredamento (mobili)

7 cloth – veiling
Schleierstoff (m)
tela (f) para velos
tissu (m) pour voiles
tessuto (m) – per veli

8 cloth – velour
Valourstoff (m)
tela (f) de terciopelado
tissu (m) panne
tessuto (m) – velour

9 cloth – velvet
Samstoff (m)
tela (f) de terciopelo
tissu-velours (m)
tessuto (m) – velluto

10 cloth – velveteen
Velvetinstoff (m)
tela (f) de velludillo
tissu (m) velours de trame
tessuto (m) – velluto di cotone

11 cloth – Venetians
Venezianerstoff (m)
tela (f) para persianas
tissu (m) pour stores
tessuto veneziano satinato

12 cloth – warehousing
Lagerstoff (m)
tela (f) para almacenes
stockage (m) de tissu
tessuto (m) – immagazzinamento

13 cloth – waste
Abfallstoff (m)
tela (f) de residuos
déchets (m pl) de tissu
tessuto (m) – cascame

14 cloth – weaving
Webstoff (m)
tela (f) tejida
tissage (m)
tessuto (m) – tessitura

15 cloth – wet finishing
Naßbahandlungstoff (m)
tela (f) rematada con agua
apprêtage (m) de tissu
tessuto (m) – finissaggio a umido

16 cloth – whipcord
Whipkordstoff (m)
paño (m) fuerte
fil (m) de fouet
tessuto (m) – corda da frusta (f)

17 cloth – white
Weißwaren (f pl)
tela (f) blanca
tissu (m) blanc
tessuto (m) – bianco

18 cloth – wool/cotton
Woll/Baumwollstoff (m)
tela (f) de lana/algodón
étoffe (f) de laine/coton
tessuto (m) – lana/cotone

1 coagulent
Koagulierungsmittel (n)
coagulante (m)
coagulant (m)
coagulante (m)

2 coarse
grob (adj)
basto, rugasa
grossier
grossolano

3 coarse count
grobe Nummer (f), grober Titer (m)
número (m) grueso
numéro (m) gros, titre (m) gros
titolo (m) grosso

4 coarse yarn
Grobgarn (n)
hilo (m) grueso
filé (m) gros
filato (m) grosso

5 coated abrasives
beschichtetes Schleifleinen (n)
abrasivos (m pl) revestidos
abrasifs (m pl) pour support
abrasivi (m pl) rivestiti

6 coated fabric
beschichtetes Gewebe (n)
tejido (m) recubierto
tissu (m) enduit
tessuto (m) rivestito

7 coating for blackout linings
Beschichtung (f) für Verdunkelungs-
stoffe (m pl)
capas (f pl) para forros de pantallas
oscuras
tissu (m) pour doublures de rideaux
rivestimenti (m pl) – per fodere da
oscuramento

8 coating for roller-blinds
Beschichtung (f) für Rollos (m pl)
capas (f pl) para persianas enrollables
tissu (m) pour stores roulants
rivestimenti (m) – per tendine ad
avvolgimento

9 coatings for thermal linings
Beschichtungen (f pl) für Thermofutter
revestimientos para forros térmicos
tissu (m) pour doublures
thermolactyles
rivestimenti (m) – per fodere termiche

10 coatings – aluminised
Beschichtungen (f pl) – mit Aluminium (f pl)
revestimientos (m pl) con aluminio
revêtements (m pl) aluminés
rivestimenti (m pl) – alluminati

11 coatings – elastomer
Beschichtungen (f pl) – aus Elastomer (f pl)
revestimientos (m pl) con fibras
elásticas
revêtements (m pl) en élastomère
rivestimenti (m pl) – elastomero

12 cobalt blue
kobaltblau (adj)
azul cobalto
bleu de cobalt
blu cobalto

13 cochineal
Koschenille
cochinilla (f)
cochenille (f)
cocciniglia (f)

14 cockled cloth
Kräuselstoff (m)
tela (f) arrugada
tissu (m) ondulé
tessuto (m) – arricciato

15 cockled yarn
Kräuselgarn (n)
hilo (m) arrugado
fil (m) ondulé
filato (m) arricciato

16 cockling
Kräuselung (f)
arrugar (m)
godage (m)
arricciatura (f)

17 cocoon
Kokon (m)
capullo (m)
cocon (m)
bozzolo (m)

18 coefficient
Koeffizient (m)
coeficiente (m)
coefficient (m)
coefficiente (m)

1 cohesion
Kohäsion (f)
cohesión (f)
cohésion (f)
coesione (f)

2 coil
Schlange (f), Windung (f)
serpentín (m), espiral (f)
serpentine (m), tour (m), spire (f)
serpentino (m), spira (f)

3 cold
kalt (adj)
frío
froid
freddo

4 cold
Kälte (f)
frío (m)
froid (m)
freddo (m)

5 cold water
Kaltwasser (n)
agua (f) fría
eau (f) froide
acqua (f) fredda

6 collateral
seitlich, Seiten–
resguardo (m)
garantie (f) additionnelle
collaterale

7 collection
Sammlung (f), Kollektion (f)
recolección (f)
récolte
raccolta (f)

8 colloid
Kolloid (n)
coloide (m)
colloîde
colloide (m)

9 colloidal
Kollodal (adj)
coloidal
colloîdal
colloidale

10 colorimeter
Kolorimeter, Farbmesser (m)
colorímetro (m)
colorimètre (m)
colorimetro (m)

11 colorimetric
kolorimetrisch (adj)
colorimétrico
colorimétrique
colorimetrico

12 colorimetry
Farbmessung (f), Kolorimetrie (f)
colorimetría (f)
colorimétrie (f)
colorimetria (f)

13 colour
Farbe (f)
color (m), matiz (m)
couleur (f)
colore (m), tinta (f)

14 colour (v)
färben
teñir, colorar
teindre, colorer
tingere

15 colouration
Farbgebung (f), Anfärbung (f)
coloración (f)
coloration (f)
colorazione (f)

16 colour blindness
Farbenblindheit (f)
daltonismo (m)
achromatopsie (f)
daltonismo (m)

17 colour card
Farbenkarte (f)
muestrario (m) de color, carta (f) de colores
carte (f) de couleurs
cartella (f) colori

18 colour change
Farbtonänderung (f), Farbenwechsel (m), Farbänderung (f)
variación (f) de tono, variación (f) de color
virage (m) de nuance, changement (m) de couleur, changement (m) de nuance
viraggio (m) di colore, variazione (f) di colore

1 colour computer matching
Computerfarbabstimmung (f)
comparacion (f) de colores por ordenador
échantillonnage (m) de couleur par ordinateur
abbinamento (m) di colori mediante computer

2 colour deviation
Farbabweichung
diferencia (f) de color
différence (f) de nuance
differenza (f) di colore

3 coloured
bunt, gefärbt, farbig (adj)
coloreado
coloré
colorato

4 coloured fibres
farbige Fasern (f pl)
fibras (f pl) coloreadas
fibres (f pl) colorées
fibre (f pl) colorate

5 coloured knop
farbige Noppe (f)
botón (m) coloreado
bouton (m) de couleur
bottone (m) colorato

6 colour fading
Farbausbleichung (f)
descolorar (m)
affaiblissement (m) de la couleur
scoloritura (f) del colore

7 colour fastness
Farbechtheit (f)
fijeza (f) de color, solidez (f) del color
solidité (f) de la couleur
solidità (f) del colore, solidità (f) tintoriale

8 colour fastness test
Farbechtheitsprüfung (f)
ensayo (m) de solidez de los colores
contrôle (m) de solidité des couleurs
prova (f) di solidità del colore

9 colouring
Farbgebung (f), Anfärbung (f)
coloración (f)
coloration (f)
colorazione (f)

10 colour kitchens
Farbküchen (f pl)
laboratorios (m pl) para comparar colores
ensemble (m) de couleurs
cucine (f pl) colorate

11 colourless
farblos (adj)
incoloro
incolore
incolore

12 colour matching
Farbabstimmung (f)
comparación (f) de colores
échantillonnage (m) de la couleur
abbinamento (m), di colori (f)

13 colour matching cabinets
Farbabstimmungsschränke
estantes para comparar colores
ensemble (m) de couleurs
armadietti (m pl) per l'abbinamento di colori

14 colour range
Farbenbereich (m)
gama (f) de colores
gamme (f) des couleurs
gamma (f) di colori

15 colour reference
Farbreferenz (f)
comparación (f) de colores
référence de couleurs
colore (m) di riferimento

16 colour reference systems
Farbbezugssysteme (n pl)
sistemas (m pl) de referencias de colores
appareils (m pl) de comparaison des teintes
sistemi (m pl) di riferimento di colori

17 colour value, dye yield, dyestuff yield
Farbstoffausbeute (f)
rendimiento (m) del colorante
rendement (m) tinctorial
resa (f) tintoriale

18 comb
Kamm (m), Blatt (n), Webeblatt (n)
peine (m)
peigne (m)
pettine (m)

1 combed cotton
gekämmte Baumwolle (f)
algodón (m) peinado
coton (m) peigné
cotone (m) pettinato

2 combed silver
gekämmtes Faserband (n)
mecha (f) peinada
ruban (m) peigné
nastro pettinato

3 combed tops
gekämmte Kammzugwickel (m pl)
husos (m pl) peinados
rubans (m pl) peignés
nastri (m pl) – top pettinati

4 combed yarn
Kammgarn (n)
hilo (m) peinado
fil (m) d'estame
filato (m) pettinato

5 combination yarn
Melangegarn (n)
hilado (m) de mezcla
filé (m) de mélange
melange, filato (m)

6 combing
Kämmen (n)
peinar (m), peinado (m)
peignage (m)
pettinatura (f), cardatura (f)

7 combing – leathers
Kammstuhl-Leder (n pl)
peinado de cueros
manchons (m pl) pour peigneuses (m)
pettinatura (f) – pelli

8 combing machine
Kämmschine (f)
máquina (f) peinadora
peigneuse (f)
pettinatrice (f)

9 combustible
feuergefährlich, brennbar (adj)
inflammable, combustible
combustible
combustibile

10 combustion
Verbrennung (f)
combustión (f)
combustion (f)
combustione (f)

11 commercial
handelsüblich, kommerziell (adj)
comercial
commercial
commerciale

12 commission
Kommission (f), Auftrag (m)
comisión (f)
commission (f)
commissione (f)

13 commission – beamers
Lohnbäumer (m pl)
tejedores (m pl) de urdimbre por encargo
ourdisseurs (m pl) – commission
commissione (f) – di insubbiamento (m)

14 commission – blenders
Kommission (f) für Mischer (m pl)
mezcladores (m pl) de encargo
mélangeurs (m pl) – commission
commissione (f) – mescolatori

15 commission – blenders of all fibres
Lohnmischer für alle Fasern (m pl)
mezcladores (m pl) por encargo de toda fibra
mélangeurs (m pl) – commission de toutes les fibres
commissione (f) – mescolatori di tutte le fibre

16 commission – burlers and menders
Lohnstopfer (m pl)
desmotadores (m pl) y remendones (m pl) por encargo
éplucheurs (m pl) et raccoutreurs (m pl) – commission
commissione (f) – e macchine da rammendo

17 commission – carders
Lohnkrempler (m pl)
cardadores (m pl) por encargo
cardeurs (m pl) – commission
commissione (f) – cardatrici

1 commission – classers
Lohnklamierer (m pl)
clasificadores (m pl) por encargo
classificateurs (m pl) – commission
commissione (f) – classificatori

2 commission – cloth making-up
Lohnkonfektion (f)
remendones (m pl) de encargo de telas, confección de géneros
présentation (f)/confection (f) de tissu – commission
commissione (f) – confezione

3 commission – cloth rerolling
Lohnarbeit für Umrollen (f)
rearrolladores (m pl) de encargo de telas
réenroulement (m) de tissu – commission
commissione (f) – riavvolgimento

4 commission – coaters
Lohnbeschichter (m pl)
revestidores (m pl) por encargo
enducteurs (m pl) – commission
commissione (f) – laminatori

5 commission – combers
Kommission (f) für Kämmer (m pl) Lohnkämmer (m pl)
cardadores (m pl) por encargo
peigneurs (m pl) – commission
commissione (f) – pettinatrici

6 commission combing
Lohnkämmerei (f)
peinado (m) por cuenta de terceros, peinado (m) por encargo
peignage (m) à façon
pettinatura (f) per conto terzi

7 commission – doublers
Lohndoublierer (m pl)
cobladores (m pl) por encargo
doubleurs (m pl) – commission
commissione (f) – binatoio (m)

8 commission – filament yarn doublers
Lohnzwirner (m pl) für Endlosgarn
dobladores (m pl) por encargo de hilacha
doubleurs (m pl) de fil continu – commission
commissione (f) – binatoio (m) filato

9 commission – finishers
Lohnappreteure (m pl)
acabadores (m pl) por encargo
apprêteurs (m pl)/finisseurs (m pl) – commission
commissione (f) – finitore (m)

10 commission finishing
Lohnausrüstung (f)
acabado (m) por cuenta de terceros
finissage (m) à façon
finissaggio (m) per conto terzi

11 commission – garnetters
Lohngarnettierer (m pl)
desmenuzadores (m pl) por encargo
garnetteurs (m pl) – commission
commissione (f) – garnettatore (m) garnettatrice (f)

12 commission – gillers
Lohnarbeit (f) für die Nadelstab-strecke (f)
peinadores (m pl) de encargo
filateurs (m pl) de gros numéros
commissione (f) – pettinatrici

13 commission – grinders
Llohnarbeit für Kratzenschleifer (f)
amoladores (m pl) de encargo, amoladores por encargo
machines (f pl) à aiguiser – commission
commissione (f) – affilatrici/ sfilacciatrice

14 commission – healders
Lohnarbeit für Weblitzenhersteller (f)
lizadores (m) de encargo
fabricants (m pl) de lisses – commission
commissione (f) – licci

15 commission – openers/synthetic
Lohnarbeit (f) für den Synthetik-Reißwolf
desmotadores (m pl) de encargo de sintético
ouvreuses (f pl) de synthétiques – commission
commissione (f) – apritoi – sintetici

16 commission – packers
Lohnpacker (m pl)
empaquetadores por encargo
emballeurs (m pl) – commission
commissione (f) – imballatori

1 commission – pattern weaving
Lohnarbeit (f) für Musterweben
tejedores sobre diseño por encargo
tissage (m) de dessins – commission
commissione (f) – tessitura di modelli

2 commission – printers
Lohndrucker (m pl)
estampadores (m pl) por encargo
imprimeurs (m pl) – commission
commissione (f) – stampatori

3 commission – pullers
Lohnarbeit (f) für Vertieher
tiradores (m pl) de encargo
boulocheurs (m pl) – commission
commissione (f) – tiratoi/strappatori

4 commission – pullers 4 cylinder La Roche
Lohnarbeit (f) für die 4 Zylinder La Roche
tiradores (m pl) de encargo – 4 cilindros La Roche
arracheuses (f pl) à 4 cylindres La Roche – à façon
commissione (f) – tiratoi – 4 cilindri La Roche

5 commission – random cutters
Lohnarbeit (f) für Gelegenheits-Zuschneider
cortadores (m pl) esporádicos por encardo
coupe (f) au hasard – commission
commissione (f) – tagliatrici casuali

6 commission – reelers
Lohnspuler (m pl)
devanadores (m pl) de encargo
dévidoirs (m pl) – commission
commissione (f) – bobinatrici

7 commission – roypacking
Lohn (m) Roypacken (n)
embalajes tipo roy por encargo
emballage (m) 'roypacking' – à façon
commissione (f) – imballaggio 'Roy'

8 commission – scourers
Lohnreiger (m pl), Lohnwäscher (m pl)
desengrasadores (m pl) de encargo
lavage (m) à fond – commission
commissione (f) – lavatrici

9 Commission – scourers – waste
Lohnarbeit (f) für Abfallwäscher
descrudadores por encargo
nettoyeurs (m pl) de déchets – à façon
commissione (f) – lavatrici – scarti

10 commission – scourers – wool
Lohnarbeit für Wollwäscher
descrudadores de residuos por encargo
dégraisseurs (m pl) de laine – à façon
commissione (f) – lavatrici – lana

11 commission – sliverers
Lohnarbeit für Kardenbandleger
descrudadores (m pl) por encargo
mèches (m pl) – commission
commissione (f) di nastri

12 commission – sorters
Kommission (f) Lohnsortierer (m pl)
clasificadores (m pl) por encargo
triage (m) – commission
commissione (f) – selezionatrici

13 commission – spinners
Lohnsprinner (m pl)
hiladores (m pl) por encargo
filateurs (m pl) – commission
filatori per conto terzi

14 commission spinning
Lohnspinnerei (f)
hilatura (f) por cuenta de terceros
filature (f) à façon
filatura (f) per conto terzi

15 commission – warpers
Lohnschärer (m pl)
urdidores (m pl) por encargo
ourdisseurs (m pl) – commission
tessitori per conto terzi

16 commission – weavers
Lohnweber (m pl)
tejedores (m pl) por encargo
tisseurs (m pl) – commission
commissione (f) – tessitori

17 commission weaving
Lohnweberei (f)
tejedura (f) por cuenta de terceros
tissage (m) à façon
tessitura (f) per conto terzi

1 commission – weft knitted fabrics
Lohnarbeit (f) für Kulierware
fabricantes (m pl) de géneros de punto de trama por encargo
tricot (m) de trame – commission
commissione (f) – maglieria in trama

2 commission – willowers
Lohnkrempler (m pl)
limpiadores (m pl) de fibras por encargo
louveteurs (m pl) – commission
commissione (f) – battitori lupo

3 commission – winders
Lohnspuler (m pl)
devanadores (m pl) de encargo
bobineurs (m pl) – commission
commissione (f) – incannatoi

4 commission – winders – cone
Lohnspuler (m pl) auf Lonen
hiladores (m pl) de conos de encargo
bobineurs (m pl) à fil croisé – à façon
commissione (f) – avvolgitrici – rocchetto conico

5 commission – winders – hand to cone
Lohnarbeit (f) von strang auf Kreuzspule
hiladores (m pl) de ovillos a conos de encargo
bobineurs (m pl) à écheveaux – à façon
commissione (f) – avvolgitrici – rocchetto conico

6 commission – yarn ballers
Lohnknäueler (m pl)
fabricantes (m pl) de ovillos por encargo
peloteuses (f pl) de fil – commission
commissione (f) – imballatrici per filati

7 commission – yarn brushers
Kommission (f) für Garnbürster (m pl) Lohn-Gernbürster (m pl)
cepilladores (m pl) de hilos por encargo
brossage (m) de fil – commission
commissione (f) – spazzolatrici per filati

8 commission – yarn doublers
Lohn – Garnfacher (m pl)
dobladores (m pl) de hilos por encargo
doubleuse (f) de fil – commission
commissione (f) – addoppiatrici per filati

9 commission – yarn hank winders
Lohn-Garnstrangdreher (m pl)
devanadores (m pl) de madejas de hilo por encargo
déméloirs (m pl) – commission
commissione (f) – avvolgitrici di matasse

10 commission – yarn heat setting
lohn Garnthemofixierer (m pl)
termofijación (f) de hilo por encargo
thermofixage (m) de fil – commission
commissione (f) termofissaggio di filato

11 commission – yarn packers
Lohn – Garnpacker (m pl)
empaquetadores (m pl) de hilo por encargo
empaquetage (m) des fils – commission
commissione (f) – imballatori del filato

12 commission – yarn reelers
Lohn – Garnhaspler (m pl)
devanadores (m pl) de hilo por encargo
bobinoirs (m pl) de fil – commission
commissione (f) – bobinatrici da filati

13 commission – yarn relaxers
lohn-Garnenspanner (m pl)
aflojadores (m pl) de hilo por encargo
relaxation (f) de fil – commission
commissione (f) – stiratori per filati

14 commission – yarn sizers
Kommission (f) für Garnnumerierer (m pl), Lohn – Garnschlichter (m pl)
encladores de hilo por encargo
encolleuses (f pl) pour filés – commission, encolleurs (m pl) des filés – à façon
commissione (f) – calibratrici da filati

15 commission – yarn twisters
Lohnzwirner (m pl)
torcedores (m pl) de hilo por encargo
retordoirs (m pl) pour filés – commission
commissione (f) – ritorcitoi di filato

16 commission – yarn winders
Lohnspuler (m pl)
hiladores (m pl) por encargo
dévideurs (m pl) à façon
commissione (f) – avvolgitrici di filati

1 common market
europäische Gemeinschaft (f)
mercado (m) común
marché (m) commun
mercato (m) comune

2 company
Firma (f)
compañía (f)
société (f)
società (f)

3 compensate
ausgleichen, kompensieren
compensar
compenser
compensare

4 compensation
Kompensation (f)
compensación (f)
compensation (f)
compensazione (f)

5 competition
Konkurrenz (f)
competencia (f)
concurrence (f)
concorrenza (f)

6 complain
klagen
reclamar
réclamer
reclamare

7 complaint
Beschwerde (f)
reclamación (f)
plainte (f)
lagnanza (f)

8 component
Komponente (f)
componente (m)
composé (m)
componente (m)

9 composition
Zusammensetzung (f), Komposition (f)
composición (f)
composition (f)
composizione (f)

10 compound
Verbindung (f)
compuesto (m)
composé (f)
composto (m)

11 compress
zusammenpressen
comprimir
compresser
comprimere

12 compressed air
Pressluft (f)
aire (m) comprimido
air (m) comprimé
aria (f) compressa

13 compressor
Verdichter (m), Kompressor (m)
compresor (m)
compresseur (m)
compressore (m)

14 comprise (v)
enthalten, einschließen
comprender, contener, incluir
comprendre
comprendere

15 compulsory
obligatorish, verpflichtend
obligatorio
obligatoire
obbligatorio

16 computer
Computer (m), Rechenmaschine (f)
ordenador (m), máquina (f) calculadora
ordinateur (m), calculatrice
computer (m), calcolatore (m)

17 computer cloth design service
Computerdesignservice (m)
sevicio (m pl) de diseños de telas por ordenador
service (m) de dessin des tissus par ordinateur
messa in carta/servizi (m pl) computerizzati di disegni in stoffa/tella

18 computer control systems for woven and knitted piece manufacturers
Computerkontrollsysteme (n pl) für Hersteller von Webe- und Wirkstückwaren

sistemas (m pl) de regulación de ordenadores para fabricantes de géneros tejidos y hechos de punto
systèmes (m pl) de commande par ordinateur pour les fabricants de pièces tissées et tricotées
sistemi (m pl) di controllo computerizzati per produttori di tessuto e maglieria

1 computer services and word processors
Computer (m pl) und Textverarbeitungs-systeme (m pl)
servicio (m pl) de ordenadores y procesado de textos
services (m pl) informatiques et machines de traitement de texte
servizi (m pl) di computer ed elaborazione di testi

2 concave
konkav (adj)
cóncavo
concave
concavo

3 concentrate
Konzentrat (n)
concentrado (m)
concentré (m)
concentrato (m)

4 concentrated
konzentriert
concentrado
concentré
concentrato

5 condensate
Kondensat (n)
líquido (m) condensado
condensat (m)
condensa (f)

6 condensation
Kondensation (f), Kondensierung (f)
condensación (f)
condensation (f)
condensazione (f)

7 condense
kondensieren, verdichten
condensar
condenser
condensare

8 condition
Zustand (m), Bedingung (f), Kondition (f)
condición (f)
condition (f)
condizione (f)

9 conditioned weight
Handelsgewicht (n)
peso (m) comercial, peso (m) acondicionado
poids (m) commercial
peso (m) commerciale

10 conditioning
Konditionieren (n), Klimatisieren (n), Konditionierung (f)
condicionar (m), acondicionamiento (m)
conditionnement (m), climatisation (f)
condizionatura (f), condizionamento (m)

11 conditioning house
Konitionierkammer (f)
departamento (m) de acondicion-amiento
condition (f) publique textile
stablimento (m) di condizionatura

12 cone
Kegel (m), Konus (m); Spule (f), konische Kreuzspule (f), Kone (f)
cono (m); bobina (f) plana, queso (m), queso (m) cónico
cône (m); bobine (f), bobine (f) conique
rocchetto (m) conico, cono (m); rocca (f), rocca (f) conica

13 cone dyeing
Konusfärben (n)
teñir (m) en cono
teinture (f) pour cônes
tintura (f) in rocca

14 cone winding
Konuswickeln (n)
devanar por conos
bobinage (m) à cône
confezione (f) dei rocchetti

15 conference
Konferenz (f)
conferencia (f)
conférence (f)
conferenza (f)

1 confidential
vertraulich (adj)
reservado
confidentiel
confidenziale

2 confirmation
Bestätigung (f)
confirmación (f), prueba
confirmation (f), épreuve (f)
conferma (f)

3 conical
konisch, kegelförmig
cónico, de forma cónica
conique
conico, a forma conica

4 coning
Konerei (f)
enconado (m)
renvidage (m) sur cônes
avvolgimento (m) su coni

5 consecutive
aufeinanderfolgend (adj)
sucesivo
successif
consecutivo/seguente

6 consider
berücksichtigen
considerar
considérer
considerare

7 considerable
erheblich, beträchtlich (adj)
considerable, notable, significante
considérable, important
considerevole, notevole

8 consignment
Lieferposten (m), Lieferung (f)
consignación (f), expedición
livraison (f)
spedizione (f)/consegna

9 consistency
Konsistenz (f)
consistencia (f)
consistance (f)
consistenza (f)

10 consist of (v)
bestehen aus
consistir en
consister en
consistere di

11 consortium
Konsortium (n)
consorcio (m)
consortium (m)
consorzio (m)

12 consultant
Berater (m)
consultante (m), especialista
expert-conseil (m)
consulente (m)

13 consultants - textile industry
Berater (m pl) – Textilindustrie (f)
expertos (m pl) industria textil
experts-conseils (m pl) dans le textile
consulenti (m pl) – industria tessile

14 contact
Berührung (f), Kontakt (m)
contacto (m)
contact (m)
contatto (m)

15 contain
enthalten
contener
contenir
contenere

16 container, vessel, beaker, can
Behälter (m), Becher (m), Gefäaß (n), Kanne (f)
recipiente (m), vaso (m), bote (m)
récipient (m), vase (m), pot (m)
recipiente (m), vaso (m)

17 container, tank, reservoir
Tank (m), Reservoir (m)
tanque (m), depósito (m)
réservoir (m)
serbatoio (m), deposito (m)

18 containerisation, containerization
verpacken in Containern (m pl)
preparar en containers
conteneurisation (f)
containerizzazione (f)

1 content
Inhalt (m), Gehalt (m)
contenido (m), tenor (m)
contenu (m), teneur (m)
contenuto (m), tenore (m)

2 continental spinning
kontinentales Spinnen (n)
hilatura (f) continental
filature (f) système continental
filatura (f) continentale

3 continuous
kontinuierlich, endlos (adj)
continuo, sin fin
continu, sans fin
continuo, senza fine

4 continuous bleaching
Dauerbleichen (n)
blanquear (m) en continuo
blanchiment (m) en continu
candeggio (m) continuo

5 continuous drying
Dauertrocknen (n)
secar (m) en continuo
séchage (m) en continu
essiccazione (f) continua

6 continuous dyeing
Dauerfärben (n), kontinuierliches Färben (n), Kontinuefärberei (f)
teñir (m) en continuo
teinture (f) en continu, teinture (f) à la continue
tintura (f) continua, tintura (f) in continuo

7 continuous dyeing machine
Kontinuefärbemaschine (f)
máquina (f) de teñido continuo
machine (f) de teinture en continu
macchina (f) per la tintura in continuo

8 continuous filament – see 'continuous yarn'

9 continuous finishing
Dauerappretur (f)
acabado continuo
finissage (m) en continu
rifinitura (f) in continuo

10 continuous scouring
Dauerreinigen (n)
descrudado continuo
débouillisage (m) en continu
lagaggio (m) in continuo

11 continuous scouring machine
Kontinue-Waschmaschine (f)
máquina (f) para descrudar en continuo
machine (f) à laver à la continue
lavatrice (f) in continuo

12 continuous spinning
Dauerspinnen (n)
hilatura (f) continua
filature (f) par continu
filatura (f) in continuo

13 continuous spinning frame
kontinuierliche Spinnmachine (f)
sistema de hilatura continua
métier (m) continu
filatoio (m) incontinuo

14 continuous washing
Dauerwaschen (n)
lavado continuo
lagage (m) en continu
lavaggio (m) incontinuo

15 continuous yarn
Endlosgarn (n)
hilo (m) continuo, filamento (m) continuo
fil (m) continu
filato (m) continuo, filo continuo, bava (f) continua

16 contra
kontra, contra
contra
contraire
contro

17 contract
Vertrag (m)
contrato (m)
contrat (m)
contratto (m)

18 contraction
Eingehen (n), Krumpfung (f), Krumpfen (n)
encogimiento (m), contracción (f)
raccourcissement (m), rétrécissement (m), retrait (m)
accorciamento (m), restringimento (m) contrazione (f)

1 contrast
Kontrast (m)
contraste (m)
contraste (m)
contrasto (m)

2 control
Kontolle (f)
mando (m)
contrôle (m)
controllo (m)

3 controlled
begrenzt (adj)
limitado
limité
controllato

4 control system
Kontrollsystem (n)
sistema (m) de control
système (m) de contrôle
sistema (m) di controllo

5 control systems
Kontrollsystem (n pl)
sistemas (m pl) de regulación
systèmes (m pl) de commande
sistemi (m pl) di controllo

6 conversion
Konversion (f), Umwandlung (f), Umrechnung (f)
conversión (f), transformación (f)
conversion (f), transformation (f)
conversione (f)

7 convert
umwandeln
convertire/transformare
convertir
convertire/trasformare

8 conveyor belt
Förderband (n)
cinta (f) transportadora
ruban (m) transporteur
nastro (m) trasportatore/cinghia trasportatrice

9 conveyor belting
Förderbandeinrichtung (f)
cintas (f pl) para el transporte
courroies (f pl) pour convoyeurs
materiale (m) per cinghie

10 conveyor systems
Förderbandsystem (n)
sistemas (m pl) de transportadores
systèmes (m pl) de bande transporteuse
sistemi (m pl) trasportatori

11 coolant
Kühlmittel (n)
refrigerante (m)
réfrigérant (m)
refrigerante (m)

12 cooling
Abkühlung (f), Erkalten (n), Kühlung (f)
refrigeración (f) enfriamiento
refroidissement (m), réfrigération (f)
raffreddamento (m), refrigerazione (f)

13 coordinate
Koordinate (f)
coordinada (f)
coordonnée
coordinata (f)

14 cop
Kops (m), Kötzer (m), Cops (m)
huso (m), husada (f), canilla (f)
canette (f), cops (m)
bobina (f), cop (m)

15 cop dyeing
Kopsfärben (n)
teñir (m) en bobina
teinture (f) sur bobine
tintura (f) in cop/in bobina

16 cop winder
Spulmaschine (f), Spulapparat (m)
canillera (f)
bobinoir (m), canetière (f)
incanntoio (m), incannatrice (f)

17 coral red
korallenrot (adj)
rojo coral
rouge corail
rosso corallo

18 cord, rope, twine
Seil (n), Strang (m), Schnur (f), Bindfaden (m)
cuerda (f), cable (m), maroma (f), cordel (m), torzal (m), bramante (m)
corde (f), boyau (m), cordon (m), ficelle (f)
corda (f), spago (m)

1 cord, rib
Rippe (f)
costilla (f)
côte (f)
costa (f)

2 corded fabrics
gerippte Gewebe (n pl)
tejidos (m pl) abordonados
tissus (m pl) côtelés
tessuti (m pl) a coste

3 cording machine
Kordelmaschine (f)
máquina (f) de fabricar cordeles
toronneuse (f)
cordatrice (f)

4 cords and twines
Schnüre (f pl) und Fäden (m pl)
cuerdas (f pl) y guitas (f pl)
cordes (f pl) et fils (m pl) retors
corde (f pl) e spaghi (m pl)/cordicelle (f pl)

5 cords – heavy industrial
fester Industriekord (m)
cuerdas (f pl) industriales de alta resistencia
cordes (f pl) pour l'industrie lourde
corde (f pl) per uso industriale

6 corduroy
Rippensamt, (m) Kord (m), Kordsamt (m)
terciopelo (m) abordonado
velours (m) côtelé
velluto (m) a coste

7 core yarn
Kerngarn (n), Coregarn (n)
hilado (m) con alma
fil (m) à âme
filato (m) con anima

8 correlation
Wechselbeziehung (f), Korrelation (f)
correlación (f)
corrélation (f)
correlazione (f)

9 correspond (v)
übereinstimmen
corresponder
correspondre
corrispondere

10 cost
Kosten (f pl)
coste (m), costo (m)
coût (m)
costo (m)

11 cost price
Kostenpreis (m), Selbstkostenpreis (m)
precio (m) de coste
frais (m pl) de revient, prix (m) de revient
prezzo (m) di costo

12 costing
Kosten (pl)
determinar el precio de coste
établissement (m) du prix de revient
costa

13 Cotswold sheep
Gloucestershire Schafe (n pl)
ovejas (f pl) de cotswolds
moutons (m pl) Cotswold
pecore (f pl) Cotswold

14 cotton
Baumwolle (f)
algodón (m)
coton (m)
cotone (m)

15 cotton and polyester spinners
Baumwoll/Polyesterspinner (m pl)
hiladores (m pl) de algodón y poliéster
filature (f) de coton et polyester
filatoi (m pl) per cotone e poliestere

16 cotton and synthetic doubling and cabling
Baumwoll- und Kunstfaser-Doublierung und -Seildrehen
doblaje y torcido de algodón y sintéticos
doublage (m) et câblage (m) de coton et synthétique
incannatura (f) accoppiamento e ritorcitura per cotone e sintetici

17 cotton blending
Baumwollmischen (f)
mezclas (m) de algodón
mélange (m) de coton
cotone (m) – mischia

1 cotton – canvas
Baumwolleinwand (f)
lona (f) de algodón
cotonnerie (f)
cotone (m) – tela

2 cotton – carding
Baumwoll – Baumwollkardieren (n)
cardar (m) algodón
cardage (m) du coton
cotone (m) – cardatura

3 cotton – combing
Baumwollkämmen (n)
peinar (m) algodón
peignage (m) du coton
cotone (m) – pettinatura

4 cotton condenser
Baumwollvlies-Verdichter (m)
condensador (m) de algodón
condenseur (m) du coton
cotone (m) – condensatore

5 cotton doublers
Baumwolldoppler (m pl),
baumwelldoublieren (m pl)
dobladores (m pl) de algodón
doubleuses (f pl) de coton
cotone (m) – accoppiatrici

6 cotton doubling
Baumwolldoppeln (n)
doblar (m) de algodón
doublage (m) de coton
cotone (m) – accoppiamento

7 cotton dyeing
Baumwollfärben (n)
teñir (m) de algodón
teinture (f) de coton
cotone (m) – tintura in cotone

8 cotton fabric
Baumwollgewebe (n)
tejido (m) de algodón
tissu (m) de coton
tessuto (m) di cotone

9 cotton fabrics
Baumwollstoffe (m pl)
tejidos (m pl) de algodón
tissus (m pl) de coton
cotone (m) – tessuti di cotone

10 cotton fibre
Baumwollfaser (f)
fibra (f) de algodón
fibre (f) de coton
fibra (f) di cotone

11 cotton gauze
Baumwollgaze (f)
gasa (f) de algodón
gaze (f) de coton
cotone (m) – garza di cotone

12 cotton grading
Baumwollklassieren (n)
clasificación (f) de algodón
classement (m) de coton
cotone (m) – classificazione di cotone

13 cottonseed
Baumwollsamen (m)
semilla de algodón
graine (f) de coton
cotone (m) – seme di cotone

14 cotton spinning
Baumwollspinnen (n),
Baumwolspinnerei (f)
hilar (m) algodón, hilatura (f) del
algodón
filature (f) du coton
cotone (m) – filatura di cotone

15 cotton system
Baumwollspinnverfahren (n)
sistema (m) de algodón
filature (f) coton
sistema (m) cotoniero

16 cotton thread
Baumwollfaden (m)
hilo (m) de algodón
fil (m) de coton
cotone (m) – filo/filato di cotone

17 cotton threads
Baumwollfäden (m pl)
hilos (m pl) de algodón
fils (m pl) de coton
cotone (m) – fili/filati di cotone

18 cotton velvet [cloth]
Baumwollsamt (m)
terciopelo (m) de algodón, pana (f) de
algodón
velours (m) de coton
velluto (m) di cotone

1 cotton weavers
Baumwollweber (m pl)
tejedores (m pl) de algodón
fabricants (m pl) de coton
cotone (m) – tessitrici di cotone

2 cotton winders
Baumwollwickler (m pl)
devanadores (m pl) de algodón
bobinoirs (m pl) de coton
cotone (m) – incannatoi di cotone

3 cotton wool
Watte (f)
algodón (m) hidrófilo
ouate (f)
cotone (m) grezzo, cotone (idrofilo)

4 cotton yarn
Baumwollgarn (n)
hilado (m) de algodón
fil (m) de coton
filato (m) di cotone

5 cotton yarn doublers
Baumwollgarndoppler (m pl)
dobladores (m pl) de hilazo de algodón
doubleuses (f pl) de fil de coton
cotone (m) – accoppiatrici

6 cotton yarn spinners
Baumwollgarnspinner (m pl)
hiladores (m pl) de algodón
filateur (m pl) de fil de coton
cotone (m) – filatori per filati

7 count
Nummer (f), Titer (m)
número (m), título (m), cuenta (f)
titre (m), numéro
titolo (m)

8 count [yarn]
Garnnummer (f)
número (m) (f) de hilos
numéro (m) [du fil]
titolo (m) [filato]

9 count (v)
auszählen
contar
compter
contare

10 counting glass
Fadenzähler (m)
cuentahilos (m)
compte-fils (m)
contafili (m)

11 count of the reed
Blattdichte (f)
cuenta (f) del peine
compte (m) du peigne
fittezza (f) del pettine

12 courtelle
Courtelle (f)
courtelle (m)
courtelle
courtelle

13 crabbing
Krabben (n)
fijar (m) tela por agua caliente,
fijación (f) en húmedo
lissage (m), crabbing (m), fixage (m)
au mouillé
avvolgimento (m) del tessuto sotto
tensione, crabbing (m) [fissaggio a
umido]

14 crabbing machine
Fiiermaschine (f)
máquina (f) fijadora del tejido
machine (f) à fixer
macchina (f) fissatrice

15 crayons – fugitives
unbeständiger Farbstift (m)
crayones (m pl) fugitivos
crayons (m pl) – fugitifs/fugaces
pastelli (m pl) – fuggitivi

16 crease
Falte (f), Bügelfalte (f)
arruga (f), doblez (m), pliegue (m),
pliegue (m) de planchado
pli (m), pli (m) de repassage
piega (f), piega (f) di stiratura

17 crease-proof
knitterfest (adj)
inarrugable
infroissable
ingualcibile

18 crease recovery
Knittererholung (f)
recuperación (f) del arrugado
défroissement (m), récupération (f)
après froissement
recupero (m) dalla gualcitura

1 crease resisting
knitterfest (adj)
resistente al arrugado
infroissable
antipiega

2 credit
Kredit (m)
crédito (m)
créance (f)
credito (m)

3 creditor
Gläubiger (m)
acreedor (m)
créancier (m)
creditore (m)

4 creel
Gestell (n), Gatter (n), Spulengatter (n), Spulengestell (n)
fileta (f)
râtelier (m), cantre (m)
rastrelliera (f), cantra (f)

5 creeling
Aufstecken (n)
devanar (m) simultáneo de un grupo de husos, colocación (f)
garnissage (m), embrochement (m)
infilamento (m), imbancamento (m)

6 creep
Schieben (n), Schlupf (m), Kriechen (n)
deslizamiento (m)
glissement (m), fluage (m)
scorrimento (m)

7 crêpe [cloth]
Krepp (m)
crepé (m), crep (m), crespón (m)
crêpe (m)
crespo (m)

8 crêpe-de-Chine [cloth]
Crêpe-de-Chine (m)
crespón (m) de China
crêpe (m) de Chine
crespo (m) di Cina

9 crêpe weave
Kreppgewebe (n), Kreppbindung (f)
tejido (m) de crespón, ligamento (m) granito
armure (f) crêpe
tessuto (m)/armatura (f) crepe

10 cretonne [cloth]
Kretonne (f)
cretona (f)
cretonne (f)
cretonne (m)

11 crimp
Kräsel (m), Kräuselung (f), Wellung (f), Ondulation (f)
rizado (m), ondulation (f)
frisure (f), ondulation (f)
increspatura (f), cretto (m), ondulazione (f)

12 crimped
gekräuselt (adj)
rizado
frisé
crettato

13 crimped fibre
Kräuselfaser (f)
fibra (f) rizada
fibre (f) frisée
fibra (f) crettata

14 crimped yarn
Kräuselgarn (n)
hilado (m) rizado
fil (m) mousse
filato (m) crettato

15 crimping
Kräuseln (n), Kräuselung (f)
cresponado (m), rizado (m) fruncido
frissage (m), ondulation (f), frisure (f), froncement (m)
arricciatura (f), crettatura (f), increspatura (f)

16 crimplene
Crimplene (f)
poliéster (m) crimplene
tissu (m) frisé/crimplène
crimplene

17 crimson
purpur (adj)
grana (f)
rouge vif
rosso sangue/cremisino

18 critical
kritisch (adj)
crítico
critique
critico

1 crock (v)
frottieren, reiben
frotar, rozar
frotter
sfregare

2 crooked
krumm, gebogen (adj)
curvo
courbé, courbe
storto/obliquo

3 crossbred wool
Crossbredwolle (f)
lana (f) de razas mezcladas, lana (f) cruzada
laine (f) croisée
lana (f) proveniente da pecore di razza incrociata, lana (f) incrociata, lana (f) crosssbred

4 cross-dye (v)
überfärben
teñir dos veces
surteindre
sovratingere

5 cross-dyeing
Überfärbung (f)
teñido bicolor
surteinture (f)
sovratintura (f)

6 crosswise
kreuzweise (adj)
cruzado
croisé
incrociato/di traverso

7 crude
roh, unbearbeitet (adj)
crudo, en bruto, en rama
écru, brut
greggio, grezzo

8 crude silk
ungekochte Seide (f), Ekrüseide (f)
seda (f) cruda
soie (f) écrue
seta (f) cruda

9 crush
Knick (m)
aplastamiento (m)
écrasement (m)
ammaccatura (f)

10 crusher
Brechmaschine (f)
máquina (f) trituradora, desbrozadora (f)
broyeuse (f)
frantoio (f), torchio (m)

11 cubic metre
Kubikmeter (m)
metro (m) cúbico
mètre cube
metro (m) cubo

12 cupro fibre
Cuprofaser (f)
fibra (f) cupro
fibre (f) cupro-ammoniacale
fibre (f) cupro

13 curl (v)
kräuseln, ringeln
cresponar, rizar
friser, onduler
arricciare

14 curtains
Vorhänge (m pl)
cortinas (f pl)
rideaux (m pl)
tende (f pl)

15 customer
Kunde (m)
cliente (m)
client (m)
cliente (m)

16 customs
Zoll (m)
aduana (f)
douane (f)
dogana (f)

17 customs declaration
Zollerklärung (f)
declaracíón (f) de aduana
déclaration (f) en douane
dichiarazione (f) doganale

18 customs duty
Zolltarif (m)
tarifa (f) aduanera/impuesto de aduana
tarif (m) douanier
tariffa (f) doganale

1 cut
Schnitt (m), Schneiden (n), Scherung (f)
corte (m)
découpage (m), coupe (f), cisaillement (m)
taglio (m)

2 cut length
Schnittlänge (f)
longitud (f) de corte
longueur (f) de coupe
lunghezza (f) di taglio

3 cut mark
Stückzeichen (n)
marca (f) de corte
marque (f) de la coupe
segno (m) del taglio, segno (m) della pezza

4 cut-pile velvet [cloth]
geschnittener Samt (m)
terciopelo (m) cortado
velours (m) coupé
velluto (m) tagliato

5 cut plush [cloth]
geschnittener Plüsch (m)
felpa (f) cortada
peluche (f) coupée
felpa (f) tagliata

6 cutting
Schnitt (m), Schneiden (n), Scherung (f)
corte (m)
découpage (m), coupe (f), cisaillement (m)
taglio (m)

7 cutting [shearing]
Scheren (n)
tundir (m)
tonte (f)
taglio (m), cimatura (f)

8 cutting [cloth]
Zuschneiden (n)
cortar (m) [tela]
tonte (f)
taglio (m) [tessuto]

9 cutting machine
Schneidemaschine (f)
máquina (f) cortadora
coupeuse (f), machine (f) à couper
taglierina (f)

10 cuttings
Stoffabfälle (m pl), Stoffabschnitte (m pl)
retales (m pl)
déchets (m pl)
ritagli (m pl)

11 cylinder
Zylinder (m)
cilindro (m)
cylindre (m)
cilindro (m)

D

12 dacron
Dacron (n)
poliéster (m) dacron
dacron (m)
dacron (m)

13 damaged
beschädigt (adj)
dañado
abimé
danneggiato

14 damask [cloth]
Damast (m)
damasco (m)
damas (m)
damasco (m)

15 damp
naß, feucht (adj)
húmedo
humide
umido

16 damping
Befeuchten (n)
humedecer (m)
mouillage (m)
umidificazione (f)

17 dangerous
schädlich, gefahrvoll, gefährlich (adj)
dañoso, nocivo, peligroso
nuisible, dangereux
dannoso, pericoloso

1 dark
dunkel, tief (adj)
oscuro
foncé
scuro

2 darning
Stopfen (n), Flicken (n)
zurcido (m)
stoppage (m), reprise (f), rentrayage (m)
rammendatura

3 darning cotton
Stopfgarn (n)
algodón (m) de coser
coton (m) à coudre
cotone (m) per rammendo

4 data
Daten (n pl)
datos (m pl)
données (f pl)
dati (m pl)

5 daylight
Tageslicht (n)
luz (f) del día
lumière (f) du jour
luce (f) diurna/di giorno

6 deadline
Termin (m), Fälligkeit (f)
límite (m) de tiempo
date (f) limite
scadenza (f) limite di tempo

7 deal
Abkommen (n), Geschäft (n)
trato (m)
affaire (f)
affare (m)

8 dealer
Händler (m)
comerciante (m)
négociant (m)
commerciante (m)

9 dealers
Händler (m pl)
comerciantes (m pl)
négociants (m pl)
commercianti (m pl)

10 dealers in second-hand machinery
Gebrauchtmaschinenhändler (m pl)
comerciantes (m pl) de maquinaria de ocasión
revendeurs (m pl) de machines d'occasion
commercianti (m pl) di macchinari usati

11 dear, expensive
teuer (adj)
caro
cher
caro

12 debit
abbuchen/Soll (n), Belastung (f)
débito (m) deuda (f)
débit (m)
debito

13 debit note
Buchungsbescheid (m) Belastungsnote (f)
nota (f) de débito
note (f) de débit
nota (f) di addebito

14 debt
Schulden (f pl)
deuda (f)
dette (f)
debito (m)

15 debtor
Schuldner (m)
deudor (m)
débiteur (m)
debitore (m)

16 decarbonize (v), decarbonise
entkarbonisieren
descarbonizar
décarboniser
decarbonizzare

17 decatising/blowing wrappers
Dekatier/Blaswickler (m)
planchadores (m pl) a vapor
intercalaires (m pl) décatissage/soufflage
fascette (f pl) per decartissaggio/soffiatura

1 dechlorination
Entchloren (n)
desclorado (m)
déchlorage (m)
declorazione (f)

2 decimal
dezimal
decimal
décimal
decimale

3 decimal system
Dezimalsystem (n)
sistema (m) decimal
système (m) décimal
sistema (m) decimale

4 decimetre
Dezimeter (m)
decímetro (m)
décimètre (m)
decimetro (m)

5 decompose (v) [decay]
zerfallen, zersetzen
descomponer, descomponerse
décomposer, se décomposer
decomporre, decomporsi

6 decompose (v) [resolve]
zerlegen, abbauen
descomponer
décomposer
scomporre

7 degree
Grad (m)
grado (m)
degré (m)
grado (m)

8 degree centigrade
Zentigrad (m)
grado (m) centigrado
degré (m) centigrade
grado (m) centigrado

9 degree [stage]
Stufe (f), Stadium (n)
grado (m), estadio (m), paso (m), etapa (f)
degré (m), stade (m), étape (m)
stadio (m)

10 decoration
Dekoration (f)
decoración (f)
décoration (f)
decorazione (f)

11 deep sea containerisation
Tiefseeverpacken (n) in Containern
empaquetado y despacho (m pl) de containers via maritima
conteneurisation (f) haute mer
containerizzazione (f) in acque profonde

12 defect
Defekt (m)
defecto (m)
défaut (m)
difetto (m)

13 defective
defekt (adj)
defectuoso
défectueux
difettoso

14 deficit
Defizit (n)
déficit (m)
déficit (m)
deficit (m)

15 degrease (v)
entfetten, entschweißen
desengrasar
dégraisser
sgrassare

16 dehaired
enthaart (adj)
pelado
ébourré
depilato

17 delay
Frist (f), Verspätung (f), Verzögerung (f)
plazo (m), retraso (m)
délai (m), retard (m)
dilazione (f), ritardo (m)

18 delicate
empfindlich (adj)
delicado
délicat
delicato

1 deliver (v)
liefern
entregar
livrer
consegnare/erogare/produrre

2 delivery
Lieferung (f)
entrega (f)
livraison (f)
consegna (f) erogazione (f) fornitura (f)

3 delivery rate
Liefergeschwindigkeit (f)
velocidad (f) de entrega
vitesse (f) de sortie, vitesse de délivraison
velocità (f) di uscita

4 delustre (v)
mattieren
deslustrar, opacar
mater
opacizzare

5 demand
Gesuch (n), Nachfrage (f), Anfrage (f), Anforderung (f)
demanda (f), exigencia (f), requisito (m)
demande (f), requête (f), exigence (f)
domanda (f), richiesta, esigenza (f)

6 denier
Denier (n)
denier (m)
denier (m)
denaro (m)

7 dense, compact, thick
dicht, kompakt, dickflüssing (adj)
denso, compacto
dense, serré
denso, fitto

8 density
Dichte (f)
densidad (f)
densité (f)
densità(f)

9 department
Abteilung (f)
departamento (m)
département
reparto (m)

10 department; service, duty
Dienst (m), Betrieb (m); Abteilung (f)
servicio (m); división (f), sección (f)
service (m)
servizio (m) divisione

11 deposit
Einlage (f), Bodensatz (m)
depósito (m)
dépôt (m)
deposito (m)

12 depot
Depot (n)
sucursal (m), almacén (m)
dépôt (m)
magazzino (m) deposito

13 depreciation
Entwertung (f), Wertminderung (f) abschreibung (f)
depreciación
dépréciation (f)
deprezzamento (m)

14 depth
Tiefe (f)
profundidad (f)
profondeur (f)
profondità

15 description
Beschreibung (f)
descripción (f)
description (f)
descrizione (f)

16 design
Design (n)
diseño (m)
dessin conception (f)
disegno (m)

17 designer
Designer (m)
diseñador (m)
dessinateur (m)
disegnatore (m) disegnatrice (f)

18 designs
Designs (n pl)
diseños (m pl)
dessins (m pl)
disegni (m pl)

1 designs – patterned textiles
Design - Textilien mit Mustern
diseños (m pl) para tejidos
dessins (m pl) – tissus imprimés
disegni (m pl) – tessili a motivi

2 despatch – see 'dispatch'

3 detailed
ausführlich, eingehend, spezifiziert (adj)
detallado
détaillé
dettagliato

4 detergents
Detergentien (n pl), Waschmittel (n pl)
detergentes (m pl)
détergents (m pl)
detergenti (m pl)

5 devaluation
Entwertung (f)
devaluación (f)
dévaluation (f)
svalutazione (f) deprezzamento

6 diagonal weave
Diagonalgewebe (n)
tejer (m) en diagonal
armure (f) diagonale
armatura diagonale

7 diagram
Diagram (n), Schaubild (n), Bild (n)
diagrama (m), gráfica (f), gráfico (m)
diagramme (m), graphique (m)
diagramma (m), grafico (m)

8 diameter
Durchmesser (m)
diámetro (m)
diamètre (m)
diametro (m)

9 different
verschieden (adj)
diferente, diverso
différent, distinct, divers
differente, diverso

10 difficult
schwer (adj)
dificil
difficile
difficile

11 dimension
Dimension (f)
dimensión (f)
dimension (f)
dimensione (f)

12 dip (sheep)
Tauchen (n)
baño (m) para ovejas
bain (m) parasiticide (pour moutons)
lavatura disinfettante (pecora)

13 direct
direkt (adj)
directo
direct
diretto

14 direct debit
Dauerauftrag (m)
débito (m) directo
prélèvement (m) bancaire
addebito (m) diretto

15 direct printing
Direktdruck (m)
estampado (m) directo
impression directe
stampa (f) diretta

16 direction [management]
Leitung (f), Direktion (f)
dirección (f)
direction (f)
direzione (f)

17 direction
Richtung (f)
sentido (m), dirección
sens (m), direction (f)
senso (m)

18 director
Direktor (m)
director (m)
directeur (m)
direttore (m)

19 directory
Telefonbuch (n), Adreßbuch (n)
guía (m)
annuaire (m)
annuario (m), guida

1 dirty
schmutzig (adj)
sucio
sale
sporco

2 discard, reject (v)
verwerfen
descartar
écarter
scartare

3 discolour
entfärben (v tr)
descolorar
déteindre
scolorare, scolorire

4 discoloration
Verfärbung (f)
descoloramiento (m)
décoloration (f)
decolorazione (f) scolorimento

5 discontinue (v)
abbrechen, unterbrechen
interrumpir
interrompre, couper
interrompere, discontinuare

6 discount
Rabatt (m)
descuento (m)
remise (f)
sconto (m)

7 dispatch (v)
spedieren, versenden, befördern, verschiffen
expedir, enviar, despachar
envoyer, expédier
spedire

8 dispatch (n)
Spedition (f), Versand (m), Verschiffung (f), Beförderung (f)
despacho (m), expedición (f), envío
expédition (f), envoi (m)
spedizione (f)

9 disperse (v)
dispergieren
dispersar
disperser
disperdere

10
displacement
Verschieben (n), Verschiebung (f)
traslación (f), desplazamiento (m)
déplacement
spostamento (m)

11 display (v)
ausstellen, aufweisen, vorführen, zeigen
exhibir, presentar
étaler, présenter
esibire, presentare, esporre

12 dissolve (v)
auflösen
disolver
dissoudre
dissolvere, sciogliere

13 distill (v)
destillieren
destilar
distiller
distillare

14 distilled water
destilliertes Wasser (n)
agua (f) destilada
eau (f) distillée
acqua (f) distillata

15 distillation
Destillation (f)
destilación (f)
distillation (f)
distillazione (f)

16 distortion
Verformung (f) Verzerrung (f), Deformation (f)
deformación
déformation
deformazione (f)

17 distribute (v)
verteilen
distribuir
distribuer
distribuire

18 distribution
Verteilung (f)
distribución (f), repartición
distribution (f), répartition
distribuzione (f), ripartizione (f)

1 diversification
Veränderung (f)
diversificacíón (f)
diversification (f)
diversificazione (f)

2 dividend
Dividende (f)
dividendo (m)
dividende (m)
dividendo (m)

3 dobby loom
Schaftstuhl (m)
telar (m) dobby
métier (m) à ratière
telaio (m) a ratiera

4 document
Dokument (n)
documento (m)
document (m)
documento (m)

5 documentation
Dokumentation (f)
documentación (f)
documentation (f)
documentazione (f)

6 doff (v)
abtrennen, abstreifen, lösen
quitar, separar, destacar
détacher, séparer, dépouiller
staccare

7 doffing
Abzug (m), Abziehen (n)
mudada (f)
levée (f)
levata (f)

8 doffing boxes
Abziehkasten (m pl)
cajas (f pl) para transportar husos
tambours (m pl) de décharge
cilindri (m pl) scaricatori

9 dogs-tooth check
Hahnentrittkaro (n)
jaquel (m) tipo diente de perro
tissu (m) pied de poule
tessuto (m) a quadretti fitti, tessuto (m) quadrettato

10 domestic market
Inlandsmarkt (m)
mercado (m) interior
marché (m) intérieur
mercato (m) interno

11 Donegal tweed cloth
Deonegal Tweed (m)
paño donegal
tweed (m) donegal
tweed (m) Donegal

12 Dorset Down sheep
Dorsetschaf (n)
oveja (f) dorset down
mouton (m) dorset down
pecora (f) del Dorset

13 Dorset Horn sheep
Dorsethornschaf (n)
oveja (f) con cuernos de dorset
mouton (m) dorset horn
pecora (f) cornuta del Dorset

14 double
doppelt (adj)
doble
double
doppio

15 doubled yarn
gefachtes Garn (n)
hilado doblado (m), hilo (m) de dos cabos
fil (m) retors à deux bouts
filato (m) binato

16 double jersey [cloth]
Doppel-Jersey (m)
jersey (m) doble
jersey (m) double
doppio jersey (m)

17 double plush [cloth]
Doppelplüsch (m)
felpa (f) doble
peluche (f) double
felpa (f) doppia

18 doubler, doubling winder
Fachmaschine (f), Dubliermaschine (f)
dobladora (f)
doubleuse (f), assembleuse (f)
binatoio (m), binatrice (f)

1 doubler, twister, doubling frame
Zwirnmachine (f)
retorcedora (f), continua (f) de retorcer
machine (f) à retordre, retordeuse (f), métier (m) à retordre
ritorcitoio (m), torcitoio (m)

2 double weave
Doppelbindung (f)
ligamento (m) doble
armure (f) double
legatura (f) doppia

3 doubling
Doublieren (n)
doblado (m)
doublage (m)
accoppiatura

4 dove grey
taubengrau (adj)
gris columbino
colombin
grigio tortora

5 drab
grau-braun, beige-braun
pardusco
drap (m) beige
grigio, grigiastro

6 draft [money]
Tratte (f)
giro (m), letra de cambio
traite (f)
tratta (f) [denaro]

7 drapery
Anzugstoffe (m pl)
género (m) para trajes
draperie (f)
drapperia (f)

8 draw in (v), suck in
saugen, ansaugen
aspirar, succionar
aspirer, sucer
aspirare, assorbire

9 draw in (v)
einziehen [Fäden]
remeter, enhebrar
remettre, aligner
passare i fili

10 dress, clothing
Anzug (m), Kleid (n)
traje (m), vestido (m), ropa (f)
costume (m), habit (m)
vestito, abito (m)

11 dress [frock]
Damenkleid (n)
vestido (m)
robe (f) pour dames
vestito da donna

12 dress, gown
Kleid (n)
traje (m), vestido (m)
vêtement, robe (f)
veste (f)

13 dress, garment, suit
Kleid (n), Anzug (m)
traje (m), vestido (m)
vêtement (m), habit (m), robe (f)
vestito (m)

14 dress (v), size
ausrüsten, appretieren
aprestar, encolar
apprêter
apprettare

15 dress (v), prepare
ansetzen, präparieren, bereiten, vorbereiten, zurichten
preparar, acabar
préparer, apprêter
preparare

16 dressing, finish
Appretur (f), Ausrüstung (f), Finish (n), Zurichtung (f), Veredlung (f)
apresto (m), acabado (m)
apprêt (m), finissage (m)
appretto (m), finissaggio (m)

17 dressing, preparation
Aufbereitung (f), Vorbehandlung (f), Bereitstellung (f), Präparation (f), Vorbereitung (f)
preparación (f)
préparation (f), apprêt (m)
preparazione (f)

18 dress material
Kleiderstoff (m)
tejido (m) para vestidos
tissu (m) pour robes
stoffa (f) per abiti

1 driving belt, drive belt
Antriebsriemen (m)
correa (f) de transmisión
courroie (f) de commande
cinghia (f) di trasmissione

2 drop wire [loom]
Fadenreiter (m), Lamelle (f)
alambre (m) de lizo
lamelle (f) casse-chaîne
lamella (f) [telaio]

3 drum
Trommel (f), Tambour (m)
tambor (m), cilindro (m)
tambour (m), cylindre (m)
tamburo (m)

4 dry
trocken (adj)
seco
sec
secco

5 dry finishing
Trockenappretur (f)
rematar en seco (m)
apprêt (m) sec
rifinitura (f) a secco

6 dryer
Trockner (m)
secador (m)
séchoir (m)
essiccatore (m)

7 drying (n)
Trocknen (n)
desecación
séchage (m)
essiccatura (f)

8 drying
Abrocknung (f)
secado
séchage (m)
asciugamento (m)

9 drying
Trocknen (n), Trocknung (f)
secado (m), secamiento (m)
séchage (m)
essiccamento (m)

10 dryness
Trockenheit (f)
sequedad (f)
sécheresse (f)
secchezza (f)

11 dry spinning
Trockenspinnen (n), Trockenspinnerei (f)
hilatura (f) en seco
filatura (f) en sec
filatura (f) a secco

12 dry weight
Trockengewicht (n)
peso (m) en seco
poids (m) sec
peso (m) secco

13 duck egg blue
enten(ei)blau (adj)
azul pálido [color de huevo de pato]
bleu d'oeuf de canard
verde-azzurro

14 ductwork
Rohrleitungsnetz (n), Verteilernetz (n)
conductos (m pl)
réseaux (m pl) de distribution – tous types
rete (f) di canali

15 dull
matt (adj)
mate, opaco
mat, opaque
opaco

16 durable
dauerhaft (adj)
durable, duradero
durable
durevole

17 duration
Dauer (f), Dauerhaftigkeit (f)
duración (f)
durée (f), vie (f) utile
durata (f)

18 dust
Staub (m)
polvo (m)
poussière (f)
polvere (f)

1 dust filtration
Staubfilterung (f)
filtración (f) de polvo
filtration (f) de poussière
filtrazione di polvere

2 dust and fume control
Staub und Rauchkontrolle (f)
regulación (f) de polvo y humo
contrôle (m) de poussière et fumée
controllo di polveree fumo

3 duty – customs
Zoll (m)
impuesto (m)
droit (m)
dazio (m) tassa (f)

4 duty, service
Dienst (m), Betrieb (m)
servicio (m)
service (m)
servizio (m)

5 duty-free
zollfrei (adj)
libre de impuestos
exempt de droits
esente da dazio, franco di dazio

6 duvetyn [cloth]
Duvetine (m)
duvetina (f)
duvetine (f)
duvettina (f)

7 dye
Farbstoff (m), Färbung (f)
tinte (m), colorante
teinture (f)
colorante

8 dyehouse carts
Wagen für Farbküchen
carros (m pl) para tintorerías
chariots (m pl) de teinturerie
carri (m pl) da tintoria

9 dye levelling agents
Farbstoffausgleicher (m pl)
agentes (m pl) igualadores de color
agents (m pl) d'unisson pour teinture
agenti (m pl) livellanti del colorante

10 dyestuff
Farbstoff (m)
colorante
matière (f) colorante
colorante

11 dyed cloth
gefärbtes Tuch (n)
tela (f) teñida
tissu (m) teint
tessuto (m) tinto

12 dyed fibre
gefärbte Faser (f)
fibra teñida
fibre (f) teinte
fibra (f) tinta

13 dyed yarn
gefärbtes Garn (n)
hilo (m) teñido
fil (m) teint
filato (m) tinto

14 dyehouse
Färberei (f)
tintorería (f)
teinturerie (f)
tintoria (f)

15 dyeing
Färben (n)
tinte (m)
teinture (f)
tintoria (f) [arte del tingere], tintura (f) [azione]

16 dyeing auxiliaries
Farbhilfsmittel (n pl)
auxiliares para teñir
auxiliaires (m pl) tinctoriaux
ausiliari (m pl) del colorante

17 dyers
Färber (m pl)
tintoreros (m pl)
teinturiers (m pl)
tinture (f pl), tintori

18 dyers – acrylic
Färber (m pl) für Acryl, Acrylfärber (m pl)
tintoreros (m pl) de géneros acrílicos
teinturiers (m pl) – acrylique
tintori (m pl) – acrilici

1 dyers – beam
Baumfärber (m pl), Kettbaumfärber (m pl)
colorantes para plegador
teinturiers (m pl) – sur l'ensouple
tintori (m pl) – subbio

2 dyers – carpet yarns
Färber (m pl) für Teppichgarn
colorantes (m pl) de hilos de alfombras
teinturiers (m pl) – fils de tapisserie
tintori (m pl) – filati per tappeti

3 dyers – cheese
Färber (m pl) für Kreuzspulen
colorantes (m pl) – bobina
teinturiers (m pl) – bobine croisée
tintori (m pl) – rocca cilindrd

4 dyers – cone
Färber (m pl) für Konus
colorantes (m pl) – cono
teinturiers (m pl) – cône
tintori (m pl) – rocca conica

5 dyers – cotton
Färber (m pl) für Baumwolle
colorantes (m pl) de algodón
teinturiers (m pl) – coton
tintori (f pl) – cotone

6 dyers – cotton and mercerised cotton
Färber (m pl) für Baumwolle und merzerisierte Baumwolle
colorantes (m pl) de algodón y algodón mercerizado (m pl)
teinturiers (m pl) – coton et coton mercerisé
tinture (f pl) – cotone e cotone mercerizzato

7 dyers – cotton and polyester
Färber (m pl) für Baumwolle und Polyester
tintoreros (m pl) de algodón y poliéster
teinturiers (m pl) – coton et polyester
tintori (f pl) – cotone e poliestere

8 dyers – cotton and polyester/cellulose blends
Färber (m pl) für Baumwolle und Polyester/Zellulosemischungen
colorantes (m pl) de algodón y mezclas de poliéster/celulosa
teinturiers (m pl) – mélanges de coton/polyester et cellulosiques
tintori (m pl) – misti poliestere/cellulosa

9 dyers – cotton yarn
Färber (m pl) für Baumwollgarn
colorantes (m pl) de hilo de algodo/n (m pl)
teinturiers (m pl) – fils de coton
tintori (m pl) – filato di cotone

10 dyers – fine count condenser yarns
Färber (m pl) für feines, Zweiaylindergarn
tintoreros (m pl) de hilos finos y condensados
teinturiers (m pl) – fils de futaine à numéro élevé
tintori (m pl) – filati condensati a titola fine

11 dyers – hair
Färber (m pl) für Haar
tintoreros (m pl) de pelo
teinturiers (m pl) – cheveux
tintori (m pl) – pelo

12 dyers – hand knitting yarns
Färber (m pl) für Handstrickgarn
colorantes hilos tejidos a hano
teinturiers (m pl) – fils de tricot à la main
tintori (m pl) – filati per lavorare a mano

13 dyers – hank
Färber (m pl) für Garnstrang
tintoreros (m pl) de madejas
teinturiers (m pl) – écheveaux
tintori (m pl) – matassa

14 dyers – hank space dyers
Färber (m pl) für (in Abständen zu färbende Garnstränge) Strang-Strecken
tintoreros (m pl) de madejas multi colores, intersticiales de ovillos
teinturiers (m pl) – écheveaux par sections
tintori (m pl) – tinture per sezioni di matasse

15 dyers – hosiery yarns
Färber (m pl) für Strumpfgarn
tintoreros (m pl) de hilos de calceteria
teinturiers (m pl) – retors pour bonneterie
tintori (m pl) – filati per calzetteria

1 dyers – loose cellulosic
Färber (m pl) für lose Zellulosefasern
tintoreros (m pl) de celulosa suelta (m pl)
teinturiers (m pl) – dérivés de cellulose en bourre
tintori (m pl) – cellulosa sciolta

2 dyers – loose cotton
Färber (m pl) für lose Baumwolle
tintoreros (m pl) de algodón suelto
teinturiers (m pl) – coton en bourre
tintori (m pl) – cotone sciolto

3 dyers – piece worsteds
Stückfärber für Kammgarn
tintoreros (m pl) de tela de estambre
teinturiers (m pl) – peignés en pièces
tintori (m pl) – pezza – pettinati

4 dyers – polyester
Färber (m pl) für Polyester
tintoreros (m pl) de poliéster
teinturiers (m pl) – polyester
tintori (m pl) – poliestere

5 dyers – polyester/linen
Färber (m pl) für Polyester/Leinen
tintoreros (m pl) de poliéster/lienzo
teinturiers (m pl) – polyester/lin
tintori (m pl) – poliestere/lino

6 dyers – polyester/viscose
Färber (m pl) für Polyester/Viskose
tintoreros (m pl) de poliéster/viscosa
teinturiers (m pl) – polyester/viscose
tintori (m pl) – poliestere/viscosa

7 dyers – polyester yarn
Färber (m pl) für Polyestergarn
tintoreros (m pl) de poliéster/hilazo
teinturiers (m pl) – fil polyester
tintori (m pl) – filato di poliestere

8 dyers – rags and waste
Färber (m pl) für Lumpen und Abfall
tintoreros (m pl) de trapos y desperdicios
teinturiers (m pl) – chiffons et déchets
tintori (m pl) – stracci e cascame

9 dyers – rayon yarn
Färber (m pl) für Reyongarn
tintoreros (m pl) de hilo de rayón
teinturiers (m pl) – fil de rayonne
tintori (m pl) – filato di raion

10 dyers – silk yarn
Färber (m pl) für Seidengarn
tintoreros (m pl) de hilo de seda
teinturiers (m pl) – soie filée
tintori (m pl) – filato di seta

11 dyers – slubbings
Färber (m pl) für Lunten
tintoreros (m pl) de mechones
teinturiers (m pl) – fil en gros
tintori (m pl) di stoppini (m pl)

12 dyers – synthetic yarns
Färber (m pl) für Synthetikgran
tintoreros (m pl) de hilos sintéticos
teinturiers (m pl) – fils synthétiques
tintori (m pl) di filati sintetici

13 dyers – tops synthetic
Färber (m pl) für Synthetik-Anzüge
tintoreros (m pl) de mechas sintéticas
teinturiers (m pl) – rubans synthétiques
tintori (m pl) di tops sintetici

14 dyers – tops wool
Färber (m pl) für Wollkammzuge
tintoreros (m pl) de rollos de lana
teinturiers (m pl) – rubans en laine
tintori (m pl) di tops di lana

15 dyers – tow
Färber (m pl) für Werg
tintoreros (m pl) de cuerdas de fibras o filamentos
teinturiers (m pl) – câbles de filaments
tintori (m pl) di capecchio

16 dyers – tow and converters
Färber (m pl) – Werggarn und Converter
tintoreros (m pl) de estopa y convertidores
teinturiers (m pl) – câble et convertisseurs
tintori (m pl) di stoppa e convertitori

1 dyers – viscose
Färber (m pl) für Viskose
tintoreros (m pl) de viscosa
teinturiers (m pl) – viscose
tintori (m pl) di viscosa

2 dyers – waste
Färber (m pl) für Abfall
tintoreros (m pl) de retales
teinturiers (m pl) – déchets
tintori (m pl) di cascame

3 dyers – wool
Färber (m pl) für Wolle
tintoreros (m pl) de lana
teinturiers (m pl) – laine
tintori (m pl) di lana

4 dyers – wool/nylon
Färber (m pl) für Wolle/Nylon
tintoreros (m pl) de lana/nilón
teinturiers (m pl) – laine/nylon
tintori (m pl) di lana/nylon

5 dyers – wool/viscose
Färber (m pl) für Wolle/Viskose
tintoreros (m pl) de lana/viscosa
teinturiers (m pl) – laine/viscose
tintori (m pl) di lana/viscosa

6 dyers – wool/polyester yarn
Färber (m pl) für Wolle/Polyestergarn
tintoreros (m pl) de hilo de lana/poliéster
teinturiers (m pl) – laine/fil de polyester
tintori (m pl) di filato lana/poliestere

7 dyers – yarn
Färber (m pl) für Garn, Garnfärber (m pl)
tintoreros (m pl) de hilado
teinturiers (m pl) – fil
tintori (m pl) di filato

8 dyers – yarn – machine knitting
Färber (m pl) für Maschinenstrickgarn
tintoreros (m pl) de hilo para máquinas de tricotar
teinturiers (m pl) – fil pour tricot à la machine
tintori (m pl) di macchina per maglieria

9 dyers – yarn – weaving
Färber (m pl) für Webgarn
tintoreros (m pl) de hilo para ser tejido
teinturiers (m pl) – fil de tissage
tintori (m pl) di filati – tessitura

10 dyers – yarn – weft knitting
Färber (m pl) für Kulierwirkgarn
tintoreros (m pl) de hilo de trama para hacer puntos
teinturiers (m pl) – fil pour tricot trame
tintori (m pl) di filati – maglieria in trama

11 dyers – yarn – woollen and worsteds
Färber (m pl) für Woll- und Kammgarn
tintoreros (m pl) de hilos de lana y estambre
teinturiers (m pl) – fils de laine et peignés
tintori (m pl) di filati – lana e pettinati

12 dyers and finishers
Färber (m pl) und Appretierer (m pl)
tintoreros (m pl) y rematadores (m pl)
teinturiers (m pl) et apprêteurs (m pl)
tintori (m pl) e finitori

13 dyers and finishers – bleachers
Färber (m pl) und Appretierer (m pl) für Bleicher
tintoreros (m pl) y rematadores (m pl) – blanqueadores
teinturiers (m pl) et apprêteurs (m pl) – blanchisseurs
tintori (m pl) e finitori – sbiancanti

14 dyers and finishers – blended
Färber (m pl) und Appretierer (m pl) für Mischungen
tintoreros (m pl) y rematadores (m pl) de mezclado
teinturiers (m pl) et apprêteurs (m pl) – mixte
tintori (m pl) e finitori – misti

15 dyers and finishers – blended and synthetic
Färber (m pl) und Appretierer (m pl) für Mischungen und Synthetiks
tintoreros (m pl) y rematadores (m pl) de mezclado y sintético
teinturiers (m pl) et apprêteurs (m pl) – mixte et synthétique
tinture (f pl) e carde finitrici – misti e sintetici

16 dyers and finishers – brushing
Färber (m pl) und Appretierer (m pl) in der Bürstenfärberei
tintoreros (m pl) y rematadores (m pl) con la brocha
teinturiers (m pl) et apprêteurs (m pl) – brossage
tintori (m pl) e finitori – spazzolatura

1 dyers and finishers – callanderers
Färber (m pl) und Appretierer (m pl) für Kalander
tintoreros (m pl) y rematadores (m pl) – aprensadores
teinturiers (m pl) et apprêteurs (m pl) – calandreurs
tintori (m pl) e finitori – calandre

2 dyers and finishers – cotton
Färber (m pl) und Appretierer (m pl) für Baumwolle
tintoreros (m pl) y rematadores (m pl) de algodón
teinturiers (m pl) et apprêteurs (m pl) – coton
tintori (m pl) e finitori – cotone

3 dyers and finishers – cotton blankets
Färber (m pl) und Appretierer (m pl) für Baumwolldecken
tintoreros (m pl) y rematadores (m pl) de mantas de algodón
teinturiers (m pl) et apprêteurs (m pl) – couvertures en coton
tintori (m pl) e finitori – coperte di cotone

4 dyers and finishers – cotton – chintzing
Färber (m pl) und Appretierer (m pl) für Baumwoll Chintz
tintoreros (m pl) y rematadores (m pl) de algodón y zaraza
teinturiers (m pl) et apprêteurs (m pl) – chintz
tintori (m pl) e finitori – cotone – lavorazione a chintz

5 dyers and finishers – cotton – fabrics
Färber (m pl) und Appretierer (m pl) für Baumwollstoffe
tintoreros (m pl) y rematadores (m pl) de géneros de algodón
teinturiers (m pl) et apprêteurs (m pl) – tissus de coton
tintori (m pl) e finitori – cotone – stoffe

6 dyers and finishers – cotton – glazing
Färber (m pl) und Appretierer (m pl) für Baumwollglanz
tintoreros (m pl) y rematadores (m pl) dando lustre al algodón
teinturiers (m pl) apprêteurs (m pl) – coton, apprêt à l'eau
tintori (m pl) e finitori – cotone – lucidatura

7 dyers and finishers – cotton/polyester
Färber (m pl) und Appretierer (m pl) für Baumwoll/Polyester
tintoreros (m pl) y rematadores (m pl) de algodón/poliéster
teinturiers (m pl) et apprêteurs (m pl) – coton/polyester
tintori (m pl) e finitori – cotone/poliestere

8 dyers and finishers – cotton/polyester rainwear
Färber (m pl) und Appretierer (m pl) für Baumwoll/Polyester Regenbekleidung
tintoreros (m pl) y rematadores (m pl) de ropa impermeable de algodón/poliéster
teinturiers (m pl) et apprêteurs (m pl) – vêtements de pluie en coton/polyester
tintori (m pl) e finitori – impermeabili cotone/poliestere

9 dyers and finishers – cotton/polyester work wear and leisure wear
Färber (m pl) und Appretierer (m pl) für Baumwoll/Polyester Arbeits- und Freizeitbekleidung
tintoreros (m pl) y rematadores (m pl) de ropa de trabajo y ropa informal hechas de algodón/poliéster
teinturiers (m pl) et apprêteurs (m pl) – vêtements de travail et de loisir en coton/polyester
tintori (m pl) e finitori – abiti da lavoro e tempo libero in cotone/poliestere

10 dyers and finishers – cotton resin finishes
Färber (m pl) und Appretierer (m pl) für Baumwollharz Finish
tintoreros (m pl) y rematadores (m pl) de algodón acabado con resina
teinturiers (m pl) et apprêteurs (m pl) – apprêts à résine de coton
tinture (f pl) e finitori – appretti resina/cotone

11 dyers and finishers – cotton/viscose blends
Färber (m pl) und Appretierer (m pl) für Baumwolle/Viskosemischungen
tintoreros (m pl) y rematadores (m pl) de algodón/viscosa
teinturiers (m pl) et apprêteurs (m pl) – apprêts à résine de coton
tintori (m pl) e finitori – misti cotone/viscosa

1 dyers and finishers – cotton – yarn
Färber (m pl) und Appretierer (m pl) für Baumwollgarn
tintoreros (m pl) y rematadores (m pl) de hilo de algodón
teinturiers (m pl) et apprêteurs (m pl) – coton filé
tintori (m pl) e finitori – filato di cotone

2 dyers and finishers – dry cleaners
Färber (m pl) und Appretierer (m pl) in – chemische Reinigungen
tintoreros (m pl) y rematadores (m pl) – limpieza en seco
teinturiers (m pl) et apprêteurs (m pl) – nettoyage à sec
tintori (m pl) e finitori – lavaggio a secco

3 dyers and finishers – flameproofers
Färber (m pl) und Appretierer (m pl) für flammenfeste Stoffe
tintoreros (m pl) y rematadores (m pl) anti inflamablea
teinturiers (m pl) et apprêteurs (m pl) – articles ininflammables
tintori (m pl) e finitori – trattamento contro incendi

4 dyers and finishers – fleece fabrics
Färber (m pl) und Appretierer (m pl) für Vliesstoffe
tintoreros (m pl) y rematadores (m pl) de géneros de vellón
teinturiers (m pl) et apprêteurs (m pl) – tissus duveteux
tintori (m pl) e finitori – tessuto di vello

5 dyers and finishers – furnishing fabrics
Färber (m pl) und Appretierer (m pl) für Möbelstoffe
tintoreros (m pl) y rematadores (m pl) de géneros para decoración
teinturiers (m pl) et apprêteurs (m pl) – tissus d'ameublement
tintori (m pl) e finitori – tessuto per arredamento

6 dyers and finishers – heat setting
Färber (m pl) und Appretierer (m pl) für Heißfixierung
tintoreros (m pl) y rematadores (m pl) termofijación
teinturiers (m pl) et apprêteurs (m pl) – thermofixage
tintori (m pl) e carde finitrici – fissaggio a caldo

7 dyers and finishers – heat setting fabrics for printing
Färber (m pl) und Veredler (m pl) – Thermofixieren von Stoffen zum Bedrucken
tintoreros (m pl) y rematadores (m pl) – termofijables al calor para estampas
teinturiers (m pl) et apprêteurs (m pl) – thermofixage des tissus pour impression
tintori (m pl) e finitori – fissaggio a caldo per stampare

8 dyers and finishers – interlinings
Färber (m pl) und Appretierer (m pl) für Zwischenfutter
tintoreros (m pl) y rematadores (m pl) para forros tiesos e internos
teinturiers (m pl) et apprêteurs (m pl) – entredoublures
tintori (m pl) e finitori – interfordere

9 dyers and finishers – knitted fabrics
Färber (m pl) und Appretierer (m pl) für Strickstoffe
tintoreros (m pl) y rematadores (m pl) de prendas hechas de punto
teinturiers (m pl) et apprêteurs (m pl) – articles tricotés
tintori (m pl) e finitori – tessuti a maglia

10 dyers and finishers – linen
Färber (m pl) und Appretierer (m pl) für Leinen
tintoreros (m pl) y rematadores (m pl) de lienzo
teinturiers (m pl) et apprêteurs (m pl) – lin
tinture (f pl) e finitori – lino

11 dyers and finishers – linen/cotton unions
Färber (m pl) und Appretierer (m pl) für Leinen/Baumwollmischungen
tintoreros (m pl) y rematadores (m pl) de lienzo/algodón unidos
teinturiers (m pl) et apprêteurs (m pl) – mélanges coton/lin
tintori (m pl) e finitori – misti lino/cotone

1 dyers and finishers – linen/polyester
Färber (m pl) und Appretierer (m pl) für Leinen/Polyester
tintoreros (m pl) y rematadores (m pl) de lienzo/poliéster
teinturiers (m pl) et apprêteurs (m pl) – lin/polyester
tintori (m pl) e finitori – lino/poliestere

2 dyers and finishers – linen/viscose
Färber (m pl) und Appretierer (m pl) für Leinen/Viskose
tintoreros (m pl) y rematadores (m pl) de lienzo /viscosa
teinturiers (m pl) et apprêteurs (m pl) – lin/viscose
tintori (m pl) e finitori – lino/viscosa

3 dyers and finishers – machinery polyester/cellulosic blends
Maschinen (f pl) für Färber (m pl) und Veredler (m pl) von Polyester/Zellularemischungen
maquinaria (f) de tintoreros y rematadores maquinaria para mezclas de poliester/cellulosa
teinturiers (m pl) et apprêteurs (m pl) – matériel pour mélanges polyester/cellulosique
tintori (m pl) e finitori – macchinari per mischie di fibra di poliestere/cellulosica

4 dyers and finishers – machinery sample lots
Maschinen (f pl) für Färber (m pl) und Veredler (m pl) von Probepartien
tintoreros y rematadores maquinaria para lotes de muestras
teinturiers (m pl) et apprêteurs (m pl) – matériel pour lots d'échantillon
tintori (m pl) e finitori – macchinari per lotti campione

5 dyers and finishers – machinery silk fabrics
Maschinen (f pl) für Färber (m pl) und Veredler (m pl) von Seidenstoffe
maquinaria (f) de tintoreros y rematadores para maquinaria para tejidos de seda
teinturiers (m pl) et apprêteurs (m pl) – matériel pour tissus de soie
tintori (m pl) e finitori – macchinari per tessuti di seta

6 dyers and finishers – machinery viscose/crease resisting
Maschinen (f pl) für Färber (m pl) und Veredler (m pl) von viskose/knitterfesten stoffen
tintoreros y rematadores para viscosa y telas maquinaria para viscosa resistente al arrugado
teinturiers (m pl) et apprêteurs (m pl) – matériel pour viscose/tissu antifroisse
tintori (m pl) e finitori – macchinari per viscosa/antipieghe

7 dyers and finishers – machinery washable wool fabrics
Maschinen (f pl) für Färber (m pl) und Veredler (m pl) von waschbaren Wollestoffen
tintoreros y rematadores maquinaria para tejidos lavables de lana
teinturiers (m pl) et apprêteurs (m pl) – matériel pour tissus de laine lavables
tintori (m pl) e finitori – tessuti di lana lavabili a macchina

8 dyers and finishers – machinery woven fabrics
Maschinen (f pl) für Färber (m pl) und Veredler (m pl) von geweben Stoffen
tintoreros y rematadores maquinaria para telas tejidas
teinturiers (m pl) et apprêteurs (m pl) – matériel pour étoffes tissées
teinturiers (m pl) et apprêteurs (m pl) tintori (m pl) e finitori – macchinari per fibre tessute

9 dyers and finishers – machinery woven nylon
Maschinen (mpl) für Färber (m pl) und Veredler (m pl) von Nylongewebe
maquinaria (f) de tintoreros y rematadores para nilón tejido
teinturiers (m pl) et apprêteurs (m pl) – matériel pour le nylon tissé
tintori (m pl) e finitori (m pl) – macchinari per nylon tessuto

10 dyers and finishers – mohair
Färber (m pl) und Appretierer (m pl) für Mohair
tintoreros (m pl) y rematadores (m pl) de angora
teinturiers (m pl) et apprêteurs (m pl) – mohair
tintori (m pl) e finitori – mohair

1 dyers and finishers – polyester
Färber (m pl) und Appretierer (m pl) für Polyester
tintoreros (m pl) y rematadores (m pl) de poliéster
teinturiers (m pl) et apprêteurs (m pl) – polyester
tintori (m pl) e finitore – poliestere

2 dyers and finishers – 100% polyester
Färber (m pl) und Appretierer (m pl) für 100% Polyester
tintoreros (m pl) y rematadores (m pl) de poliéster puro (100%)
teinturiers (m pl) et apprêteurs (m pl) – 100% polyester
tintori (m pl) e finitori – 100% poliestere

3 dyers and finishers – polyester/cotton blends
Färber (m pl) und Appretierer (m pl) für Polyester/Baumwollmischungen
tintoreros (m pl) y rematadores (m pl) de mezclas de poliéster/algodón
teinturiers (m pl) et apprêteurs (m pl) – mélanges polyester/viscose
tintori (m pl) e finitori – misti poliestere/cotone

4 dyers and finishers – polyester/viscose blends
Färber (m pl) und Appretierer (m pl) für Polyester/Viskosemischungen
tintoreros (m pl) y rematadores (m pl) de mezclas de poliéster/viscosa
teinturiers (m pl) et apprêteurs (m pl) – mélanges polyester/viscose
tintori (m pl) e finitori – misti poliestere/viscosa

5 dyers and finishers – printers
Färber (m pl) und Appretierer (m pl) für Druckstoffe
tintoreros (m pl) y rematadores (m pl) – estampadores
teinturiers (m pl) et apprêteurs (m pl) – imprimeurs
tintori (m pl) e finitori – stampatori

6 dyers and finishers – proofers
Färber (m pl) und Appretierer (m pl) für Imprägnierer
tintoreros (m pl) y rematadores (m pl) – tejidos
teinturiers (m pl) et apprêteurs (m pl) – articles imperméabilisés
tintori (m pl) e finitori – tessuti impermeabili

7 dyers and finishers – raising
Färber (m pl) und Appretierer (m pl) für Aufrauher
tintoreros (m pl) y rematadores (m pl) – perchadores
teinturiers (m pl) et apprêteurs (m pl) – grattage
tintori (m pl) e carde finitrici – garzatura

8 dyers and finishers – rot proofers
Färber (m pl) und Appretierer (m pl) für fäulnisbeständige Textilien
tintoreros (m pl) y rematadores (m pl) resistentes a la putrefacción
teinturiers (m pl) et apprêteurs (m pl) – protection contre la pourriture
tintori (m pl) e finitori – tessuti resistenti alla putrefazione

9 dyers and finishers – scourers and millers
Färber (m pl) und Appretierer (m pl) für Waschen und Walken
tintoreros (m pl) y rematadores (m pl) – batanados
teinturiers (m pl) et apprêteurs (m pl) – décreuseurs et fouleurs
tintori (m pl) e finitori – per lavatrici e follatrici

10 dyers and finishers – showerprooofers
Färber (m pl) und Appretierer (m pl) für Duschimprägnierer
tintoreros (m pl) y rematadores (m pl) de impermeabilizantes
teinturiers (m pl) et apprêteurs (m pl) – hydrofuge
tintori (m pl) e finitori – tessuti impermeabili

11 dyers and finishers – shrinkers
Färber (m pl) und Appretierer (m pl) gegen Schrümpf
tintoreros (m pl) y rematadores (m pl) de encogidos
teinturiers (m pl) et apprêteurs (m pl) – rétrécissement
tintori (m pl) e finitori – tessuti irrestringibili

12 dyers and finishers – synthetics
Färber (m pl) und Appretierer (m pl) für Synthetiks
tintoreros (m pl) y rematadores (m mp) de sintéticos
teinturiers (m pl) et apprêteurs (m pl) – synthétiques
tintori (m pl) e finitori – sintetici

1 dyers and finishers – teazle gigged finishes
Färber (m pl) und Appretierer (m pl) für distelgerauhtes Finish
tintoreros (m pl) y rematadores (m pl) – remate de carda tundida
teinturiers (m pl) et apprêteurs (m pl) – aprêts du grattage
tintori (m pl) e finitori – appretti garzati

2 dyers and finishers – terry fabrics
Färber (m pl) und Appretierer (m pl) für Frottierstoffe
tintoreros (m pl) y rematadores (m pl) de géneros de rizo
teinturiers (m pl) et apprêteurs (m pl) – tissus éponge
tintori (m pl) e finitori – tessuti a spugna

3 dyers and finishers – tubular knitted fabrics
Färber (m pl) und Appretierer (m pl) für Schlauchstrickware
tintoreros (m pl) y rematadores (m pl) de géneros de punto
teinturiers (m pl) et apprêteurs (m pl) – tissus tricotés tubulaires
tintori (m pl) e finitori – tessuti tubolari

4 dyers and finishers – viscose
Färber (m pl) und Appretierer (m pl) für Viskose
tintoreros (m pl) y rematadores (m pl) de viscosa
teinturiers (m pl) et apprêteurs (m pl) – viscose
tintori (m pl) e finitori – viscosa

5 dyers and finishers – viscose/cotton blends
Färber (m pl) und Appretierer (m pl) für Viskose/Baumwollmischungen
tintoreros (m pl) y rematadores (m pl) de mezclas de viscosa/algodón
teinturiers (m pl) et apprêteurs (m pl) – mélanges coton/viscose
tintori (m pl) e finitori – misti viscosa/cotone

6 dyers and finishers – viscose/flax
Färber (m pl) und Appretierer (m pl) für Viskose/Leinen
tintoreros (m pl) y rematadores (m pl) de viscosa/lino
teinturiers (m pl) et apprêteurs (m pl) – viscose/lin
tintori (m pl) e finitori – viscosa/fibra di lino

7 dyers and finishers – viscose/polyester
Färber (m pl) und Appretierer (m pl) für Viskose/Polyester
tintoreros (m pl) y rematadores (m pl) de viscosa/poliéster
teinturiers (m pl) et apprêteurs (m pl) – viscose/polyester
tintori (m pl) e finitori – viscosa/poliestere

8 dyers and finishers – viscose/polyester and flax
Färber (m pl) und Appretierer (m pl) für Viskose/Polyester und Leinen
tintoreros (m pl) y rematadores (m pl) de viscosa/poliéster y lino
teinturiers (m pl) et apprêteurs (m pl) – viscose/polyester et lin
tintori (m pl) e finitori – viscosa/poliestere e fibra di lino

9 dyers and finishers – waterproofers
Färber (m pl) und Appretierer (m pl) für Wasserimprägnierer
tintoreros (m pl) y rematadores (m pl) de impermeabilizantes
teinturiers (m pl) et apprêteurs (m pl) – imperméabilisation
tintori (m pl) e finitori – tessuti impermeabili

10 dyers and finishers – wool/cotton union
Färber (m pl) und Appretierer (m pl) für Wolle/Baumwollmischungen
tintoreros (m pl) y rematadores (m pl) de lana/algodón unidos
teinturiers (m pl) et apprêteurs (m pl) – mélange laine/coton
tintori (m pl) e finitori – misto lana/cotone

11 dyers and finishers – woollen/worsted
Färber (m pl) und Appretierer (m pl) für Stoffe aus Streichgarn und Kamgarn
tintoreros (m pl) y rematadores (m pl) de lana/estambre
teinturiers (m pl) et apprêteurs (m pl) – lainage/peigné
tintori (m pl) e finitori – lana/pettinati

12 dyers and finishers – wool/polyester
Färber (m pl) und Appretierer (m pl) für Wolle/polyester
tintoreros (m pl) y rematadores (m pl) de lana/Poliéster
teinturiers (m pl) et apprêteurs (m pl) – laine/polyester
tintori (m pl) e finitori – lana/poliestere

1 dyers and finishers – workwear and leisurewear
Färber (m pl) und Appretierer (m pl) für Arbeits- und Freizeitbekleidung
tintoreros (m pl) y rematadores (m pl) de ropa de trabajo y ropa informal
teinturiers (m pl) et apprêteurs (m pl) – vêtements de travail et de loisir
tintori (m pl) e finitori – abbigliamento da lavoro e per il tempo libero

2 dyes
Farbstoffe (m pl)
tintes (m pl)
teintures (f pl)
coloranti

3 dyes and pigments for textile printing
Farbstoffe (m pl) und Pigmente (n pl) für Textildruck, Färbemittel und Pigments für Textildrucke
tintes (f pl) y pigmentos para estampar tejidos
teintures (f pl) et pigments (f pl) pour l'impression des textiles/tissus
coloranti (f pl) e pigmenti per stampa tessile, tinture e pigmenti per stampaggio tessile

4 dyes for cellulosic
Farbstoffe (m pl) für Zellulose
tintes (f pl) para celulosa
teintures (f pl) pour dérivés du cellulose
tinture (f pl) per sostanze cellulosiche

5 dyes for leather
Farbstoffe (m pl) für Leder
tintes (f pl) para cuero
teintures (f pl) pour cuir
coloranti (f pl) per pelli

6 dyes for synthetic
Farbstoffe (m pl) für Synthetiks
tintes (f pl) para sintético
teintures (f pl) pour synthétiques
tinture (f pl) per sostanze sintetiche

7 dyes for transfer-printing wool and wool blends
Farbstoffe (m pl) für Transferdruck auf Wolle und Wollmischungen
tintes para la estampación de lana y mezclas de lana por transferencia
teintures (f pl) pour impression par copie hectographique, laine et mélange de laine
coloranti (f pl) per trasferimento di stampa in lana/misto lana

8 dyes for wool
Farbstoffe (m pl) für Wolle
tintes (f pl) para lana
teintures (f pl) pour laine
coloranti (f pl) per lana

9 dyestuffs
Farbstoffe (m pl)
tintes (f pl)
matières (f) colorantes
coloranti (m pl)

E

10 edge
Kante (f), Rand (m)
filo (m)/orillo
bord (m)
bordo (m)

11 education
Ausbildung (f)
enseñanza (f)
enseignement (m)
istruzione (f), educazione

12 effluent treatment
Abwasserbehandlung (f)
tratamiento (m) efluvente
traitement (m) de l'eau résiduaire
trattamento (m) delle acque di rifiuto

13 elastic
elastisch (adj)
elástico
élastique
elastico

14 elasticity
Elastizität (f)
elasticidad (f)
élasticité (f)
elasticità (f)

15 electric
elektrisch (adj)
eléctrico
électrique
elettrico

1 electrical engineers and suppliers
Elektrotechniker (m pl) und Lieferanten (m pl)
suministradores (m pl) y electricistas (m pl)
ingénieurs (m pl) électriciens et fournisseurs
ingegneri (m pl) elettrici e fornitori

2 electricity
Elektrizität (f)
electricidad (f)
électricité (f)
elettricità (f)

3 electromagnetic brakes, D.C.
elektromagnetische Bremsen (f pl) für Gleichstrom
frenos (m pl) electromagnéticos, corriente continua
freins (m pl) électromagnétiques C.C.
cavalletti (m pl) elettromagnetici a corrente continua

4 electromagnetic clutches, D.C.
elektromagnetische Kupplungen (f pl) für Gleichstrom
embragues (m pl) electromagnéticos, corriente continua
embrayages (m pl) électromagnétiques C.C.
innesti (m pl) electromagnetici a corrente continua

5 electronic
Elektronen-
electrónico
électronique
elettronico

6 electronic textile equipment
Textilelektronik (f)
equipos (m pl) electrónicos para tejidos
équipement (m) électronique textile
attrezzatura (f) tessile elettronica

7 embossing
Gaufrieren (n)
realzar (m)
gaufrage (m)
goffratura (f)

8 emulsions
Emulsionen (f pl)
emulsiones (f pl)
émulsion (f pl)
emulsioni (f pl)

9 end spacing
Kettfäden (m pl) in gleichen Abständen
separación (f) terminal de urdimbre
écartement (m) des fils de chaîne
densità (f) dei fili

10 engineering speciality oils
technische Spezialöle (m pl)
aceites (m pl) especiales para talleres
huiles (f pl) spéciales de mécanique
oli (m pl) speciali per l'ingegneria

11 engineers and millwrights
Ingenieure (m pl) und Maschinenschlosser (m pl)
ingenieros (m pl) y constructores (m pl) de talleres
ingénieurs (m pl) et constructeurs (m pl) de moulins
ingegneri (m pl) e fresatori

12 evaporate
verdampfen, verdunsten
evaporar
évaporer
evaporare

13 examiner wool, yarns, cloths
Frobewolle (f), -fäden (m pl), -stoffe (m pl)
lana, hilos y telas controlados
échantillons (m pl) de laine, fils et tissus
esaminatore di lana, filati, tessuti

14 export
Export (m), Ausfuhr (f), exportieren (v)
exportación (f)
exporter (v)
esportare (v)

15 exporters
Exporteur (m pl), Ausfuhrhändler (m pl)
exportadores (m pl)
exportateurs (m pl)
esportatori (m pl)

1 export documentation
Exportdokumentation (f)
documentación (f) de exportación
documentation (f) d'exportation
documentazione per export

2 export purchasing agents
Exporteinkäufer
agentes (m pl) de compras para la exportación
agents (m pl) d'achat d'exportation
agenti (m pl) addetti all'acquisto di prodotti d'esportazione

3 extraction
Auszug (m)
extracción (f)
extraction (f)
estrazione (f)

4 extraction, waste
Extraktionsabfall (m)
residuos (m pl) extraídos
déchets (m pl) d'extraction
estrazione di cascame/scarto

5 extrusion
Strangpressen (n), Extrusion (f)
estiramiento (m) por presión
extrusion (f)
estrusione (f)

F

6 fabric – batching units
Stoff (m) – Chargeneinheiten (f pl) von Stoff
tejido (m) – unidades de lote
unités (f pl) d'enroulage du tissu
tessuto (m) – unità di oliaggio

7 fabric bedspreads
Gewebebettbezüge (m pl)
tela (f) para colchas
dessus (m pl) de lit en tissu
stoffa (f) per copriletti

8 fabric coating
beschichtetes Gewebe (n)
bayetón (m)
revêtement (m) tissu
rivestimento di tessuto

9 fabric contract
Uniform (f) – Gewebe (m)
tela (f) de contracción
tissu (m) par contrat
stoffa (f) – contratto

10 fabric – co-ordinates
Stoff (m) – Kombinationen (f pl)
telas (f pl) coordenadas
tissus (m pl) coordonnés
tessuto (m) – coordinati

11 fabric – crease resisting – cotton, viscose and blends
Stoff (m) – knitterresistent aus Baumwolle, Viskose und Mischungen
telas (f pl) resistentes a arrugarse – algodón, viscosa y mezclas
tissu (m) antifroisse de coton, viscose et mélanges
tessuto (m) – ingualcibile – cotone, viscosa e misti fibre

12 fabric, curtaining
Verhangstoff (m)
tela (f) para cotrinas
tissu (m) pour irdeaux
stoffa (f) per tendaggi

13 fabric, dress
Kleiderstoff (m)
género (m) para vestidos
étoffe (f) d'habillement
stoffa (f) per vestiti

14 fabric, elastic
elastisches Gewebe (n)
tela (f) elástica
élastique (m) pour tissu
elastico (m) per stoffe

15 fabric – embroidery
Stoff (m) – Stickerei (f), Stickereistoff (m)
telas (f pl) bordadas
tissu (m) de broderie
tessuto (m) – da ricamo

16 fabric – fire-resistant
feuerfestes Gewebe (n)
tela (f) resistente al fuego
tissu (m) ininflammable
tessuto (m) – resistente al fuoco

1 fabric – flameproof
flammenfestes Gewebe (n)
tela (f) resistente a las lanas
tissu (m) – ininflammable
tessuto (m) inifiammabile

2 fabric – furnishing
Dekorationsstoff (m)
tela (f) para decoración
tissu (m) pour la maison
stoffa (f) per arredamenti

3 fabric – leisurewear
Freizeitbekleidungstoff (m)
tela (f) para ropas informal
tissu (m) pour vêtements de loisir
stoffa (f) – abbigliamento leggero

4 fabric – lingerie
Gewebe (n) für Damenwäsche
tela (f) para ropa interiores y
femeninas
tissu (m) pour lingerie
stoffa (f) per biancheria intima

5 fabric – paper and board
machines
Stoff (m) für Papier- und
Pappe-Maschinen
telas (f pl) para maquinaria de
papel y cartón
machines (f pl) à papier et à
carton
tessuto (m) – macchine per carta
e cartone

6 fabric – parachute
Fallschirmgewebe (n)
tela (f) para paracaídas
tissu (m) pour parachutes
stoffa (f) per paracaduti

7 fabric – polipropileno
Polypropylengewebe (n)
tela (f) de propileno
tissu (m) polypropylène
stoffa (f) – polipropilene

8 fabric – printers' blankets
Stoff (m) – Druckmitläufer,
Drucktuch (n)
telas (f pl) para cubiertas de
estampación
toiles (f pl) d'impression
tessuto (m) – coperte per
stampaggio

9 fabric – printers
Gewebedrucker (m pl)
tela (f) para estampación
imprimeurs (m pl) des tissus
stoffe (f pl) per stampa

10 fabric – silk
Stoff (m) – Seide (f), Seidenstoff (m)
tela (f) de seda
tissu (m) de soie
tessuto (m) – seta

11 fabric softeners
Weichmacher (m pl) [in
Waschmitteln]
sauvizantes (m pl) de tela
porduits (m pl) assouplissants
ammorbidenti (m pl) per stoffe

12 fabric, spinnaker
Spinnaker-Stoff (m)
tela (f) para velas de yates
voile (f) ballon
stoffa (f) – spinnaker

13 fabric, upholstery
Möbelstoff (m)
tela (f) para tapiceria
tissu (m) d'ameublement
stoffa (f) per tappezzeria

14 factory removals
Betriebsabbau (m)
transportes de la fábrica
elimination (f) en usine/
transport (m)
trasportatori di fabbrica

15 felling [weaving]
Stückabschluß (m) [Weben]
remate (m)
rabattage (m)
sfilacciatura (f) degli orli
[tessitura]

16 felting
Filzbildung (f)
fieltro (m)
feutrage (m)
feltratura (f)

17 fibre
Faser (f)
fibra (f)
fibre (f)
fibra (f)

1 fibre – fillings
Füllfaser (f)
relleno (m) de fibras
fibre (f) de rembourrage
fibra (f) – sostanze per date consistenza e peso; filo che va da una cimosa all'altra nell'orditura; carica, alimentazione [carda]

2 fibre – fillings – toys and soft furnishings
Füllfaser für – Spielzeug (m) und weiche Heimtextilien (f pl)
fibras (f pl) para rellenar juguetes y muebles
fibre (f) de rembourrage pour jouets et ameublement
fibra (f) – per imbottiture – giocattoli ed arredamenti in tessuto

3 fibre – flame retardent
flammenfeste Faser (f)
fibra (f) resistente al fuego
fibre (f) ignifuge
fibra (f) – antifiamma

4 fibre – lubricants
Schmälzen (f pl) für Faser
fibra (f) de lubrificantes
lubrifiants (m pl) pour fibre
fibra (f) – lubrificanti

5 fibre – marketing
Marketing/Förderung (f) für Faser
fibra (f) comercial
commercialisation (f) des fibres
fibra (f) – commerciale

6 fibre – natural
Naturfaser (f)
fibra (f) natural
fibre (f) naturelle
fibra (f) – naturale

7 fibre – nylon
Faser (f) – Nylon (n), Nylonfaser (f)
fibras (f pl) de nilón
fibre (f) de nylon
fibra (f) – nailon

8 fibre – polyamide
Faser (f) – Polyamid (m), Polymidfaser (f)
fibras (f pl) de poliamida
fibre (f) de polyamide
fibra (f) – poliammide

9 fibre – polyester
Faser (f) – Polyester (m), Polyester faser (f)
fibras (f pl) de poliéster
fibre (f) de polyester
fibra (f) – poliestere

10 fibre – polypropylene
Faser (f) – Polypropylen (m), Polyproylenfaser (f)
fibras (f pl) de polipropileno
fibre (f) de polypropylène
fibra (f) – polipropilene

11 fibre – rayon
Faser (f) – Reyon, Rayfaser
fibras (f pl) de rayón
fibre (f) de rayonne
fibra (f) – raion

12 fibre – reclaimers
Wiederverwerter (m) von Faser
recuperadores (m pl) de fibras
récupérateurs (m pl) de fibre
fibra (f) – rigeneratori

13 fibre – speciality
Spezialfasern (f pl)
especialidad (f) de fibras
fibre (f) spéciale
fibra (f) – speciale/specialità

14 fibre staple – nylon
Spinnfaser (f) – Nylon (n)
filamentos (m pl) de fibras de nilón
fibre (f) de nylon coupée
fibra (f) fiocca – nylon

15 fibre staple – polyester
Spinnfaser (f) – Polyester (m)
fibra (f) cortada-poliester
fibre (f) de polyester coupée
fibra (f) fiocca – poliestere

16 fibre – substandard
Substandardfaser (f)
fibra (f) de baja calidad
fibre (f) de qualité inférieure
fibra (f) – scadente

17 fibre – synthetic
Chemiefaser (f)
fibra (f) sintética
fibre (f) synthétique
fibra (f) – sintetica

1 fibre – viscose
Viskosefaser (f)
fibra (f) de viscosa
fibre (f) de viscose
fibra (f) – viscosa

2 filament – viscose
Viskoseendlosfaser (f)
filamento (m) de viscosa
filement (m) de viscose
filamento (m) – viscosa

3 fine animal fur
feines Fell (n)
piel (f) fina de animales
poil (m) animal fin
pelliccia (f) di animale dal pelo fino

4 finished width
Fertigbreite (f)
anchura (f) final
largeur (f) finale
larghezza (f) finita

5 flame retardant agent
flammenfestes Mittel (n)
agente (m) resistente al fuego
agent (m) ignifuge
agente (m) antifiamma

6 flat knitting
Flachstrickerei (f)
puntos (m pl) planos
tricotage (m) rectiligne
lavorare (v) a maglie diritte; maglieria rettilinea

7 flax
Flachs (m)
lino (m)
lin (m)
fibra (f) di lino

8 flax/linen
Flachs (m)/Leinen (n)
lino (m)/tejido de lino
lin (m)/tissu (m) de lin
fibra (f) di lino/lino

9 flock manufacturers
Flockehersteller (m pl)
fabricantes (m pl) de borra
fabricants (m pl) de floc
produttori (m pl) di barra (f) biaccolo (m)

10 flock and rag manufacturers
Flocke- und Lumpenhändler (m pl)
fabricantes (m pl) de borra y trapos
négociants (m pl) de floc et de chiffon
produttori (m pl) di borra e stracci

11 fluid
Flüssigkeit (f)
fluido (m)
fluide (m)
fluido (m)

12 fluorescent brightening agents
fluoreszierende Aufheller (n pl)
agentes (m pl) fluorescentes para resaltar
agents (m pl) d'avivage fluorescents
agenti (m pl) ravvivanti e fluorescenti

13 freight forwarders
Güterspediteure (m pl) Spediteure (m pl)
transitarios, transportistas
transitaires (m pl)
spedizionieri (m pl)

14 freight forwarders – air
Luftfrachtexpediteur (m), Luftfrachtspediteur (m)
transportistas – via aérea
transitaires (m pl) par voie d'avion
spedizionieri (m pl) – via aerea

15 freight forwarders – deep sea
seefrachtspediteur (m)
transportistas – via marítima
transitaires (m pl) par voie maritime
spedizionieri (m pl) – via mare

16 fur fabric manufacturers
Fellhersteller (m pl)
fabricantes (m pl) de telas de piel
fabricants (m pl) de fourrure synthétique
produttori (m pl) di stoffe per pellicce

1 furnishing brocades
Dekorationsbrokate (m pl)
brocados (m pl) para decoración
brocart (m) pour garniture de siège
broccati (m pl) per arredamenti

G

2 gas
Gas (n)
gas (m)
gaz (m)
gas (m)

3 gauge
Gauge (n), Meßgerät (n)
manómetro (m)
jauge (f)
calibro (m)

4 garnetters
Garnettierer (m pl)
desmenuzadores (m pl) de telas
garnetteurs (m pl)
garnettatrici (f pl)

5 garnetting
Fadenöffnen (n)
desmenuzar (m) de telas
garnettage (m)
garnettatura (f)

6 gilling
Nadelstabbehandeln (n)
peinadura (f)
gillsage (m)
pettinatrici (f pl)

7 glass
Glas (n)
vidrio (m)
verre (m)
vetro (m)

8 glaze
Glanz (m)
vidriar (v)
aspect (m) glace
lucentezza (f), lucido (m)

9 glycol
Glykol (m)
glicol (m)
glycol (m)
glicole (m)

10 goods
Ware (f)
mercancía
marchandise (f)
merci (f)

11 grading
Abstufung (f)
clasificación (f)
évaluation (f) du grade
classificazione (f)

12 grammes
Gramm (n)
gramos (m pl)
grammes (m pl)
grammi (m pl)

13 greasy
fett/fettig (adj)
con grasa
graisseux
unto

14 greasy cloth
Schmutzstoff (m)
tela (f) con grasa
tissu (m) graisseux
tessuto (m) unto

15 greasy wool
Schmutzwolle (f)
lana (f) con grasa
laine (f) en suint
lana (f) grassa

16 greasy yarn
Schmutzgarn
hilo (m) con grasa
fil (m) graisseux
filato (m) grasso

17 grey
grau (adj)
gris
gris
grigio

1 grey-cloth
Rohware (f)
tejido (m) en crudo
tissu (m) écru
tessuto greggio

2 gripper loom
Greifer-Webstuhl (m)
telar (m) con pinzas
métier à tisser à pince
telaio (m) a pinza

3 guarantee
Garantie (f)
garantía (f)
garantie (f)
garanzia (f)

4 guide
führen (v)
guía (f)
guide (m)
guida (f)

H

5 haircord [cloth]
Haarcord (m)
paño (m) de pelo
tapis (m) en poil animal
corda (f) di crine

6 hairs – alpaca
Alpakahaar (n)
pelos (m) de angora
poil (m) de chèvre d'angora
pelo (m) – angora

7 hairs – angora rabbit
Angorakaninchen (n)
pelos (m pl) de conejo de angora
poil (m) de lapin d'angora
pelo (m) – coniglio d'angora

8 hairs – camel
Kamelhaar (n)
pelo (m) de camello
poil (m) de chameau
pelo (m) – cammello

9 hairs – camel-combers
Kamelhaarkämmer (m pl)
pelo (m) – peinados de camello
peigneurs (m pl) de poil de chameau
pelo (m) – cammello – pettinatrici

10 hairs – camel, dehaired
Kamel (n) – enthaart
pelo (m) de camello sin pelo
poil (m) de chameau ébourré
pelo (m) – cammello – depilato

11 hairs – camel, scoured
Kamel (n) – gereinigt
pelo (m) desengrasado de camello
poil (m) de chameau lavé à fond
pelo (m) – cammello – lavato

12 hairs – cashmere
Kaschmirhaar (n)
pelos (m pl) de cachemir
poil (m) de cachemire
pelo (m) – cashmere

13 hairs – cashmere by-products
Haare (n pl) – Kaschmir
Nebenprodukte
pelos (m pl) derivados de cachemir
poils (m pl) – sous-produits cachemire
pelo (m) – sottoprodotti in cashmere

14 hairs – cashmere combers
Haare (n pl) – Kaschmir-Kämmer
pelos (m pl) del peinar de cachemir
poils (m pl) – peigneurs de cachemire
pelo (m) – cashmere – pettinatrici

15 hairs – cashmere products
Haare (n pl) für Kaschmirprodukte (n pl)
productos (m pl) de pelos de cachemir
articles (f pl) en poil de cachemire
pelo (m) – prodotti in cashmere

16 hairs – cashmere, dehaired
Kaschmir (m) – enthaart
pelos (m pl) cachemir sin pelo
poil (m) de cachemire ébourré
pelo (m) – cashmere – depilato

17 hairs – cashmere raw
Rohkaschmir (m)
pelos (m pl) crudos de cachemir
poil (m) de cachemire brut
pelo (m) – cashmere grezzo

18 hairs – cashmere scoured
Kaschmir (m) – gereinigt
pelos (m pl) descrudado de cachemir
poil (m) de cachemire lavé a fond
pelo (m) – cashmere – lavato

1 hairs – dehaired
enthaart (adj)
pelo (m pl) pelado
poil (m) ébourré
pelo (m) – depilato

2 hair – dehaired Mongolian and Chinese cashmere
mongolischer und chinesischer Kaschmir (m) – enthaart
pelo (m pl) cachemir mongólico y china pelado
poil (m) de cachemire mongolien et chinois ébourré
pelo (m) – cashmere mongolo e cinese – depilato

3 hairs – depigmented
Entpigmentieren (n)
pelo (m) sin colorante
poil (m) dépigmenté
pelo (m) – depigmentato

4 hairs – exporter
Exporteur (m) [von Haar]
pelo (m pl) de exportación
exportateur (m) de poil
pelo (m) di esportatore

5 hairs – goat
Ziegenhaar (n)
pelos (m pl) de cabra
poil (m) de chèvre
pelo (m) di capra

6 hairs – Guanaco
Guanakohaar (n)
pelo (m) de guanaco
poil (m) de guanaco
pelo (m) di guanaco

7 hairs – human
Menschenhaar (n)
pelo (m) humano
cheveux (m pl)
pelo (m) – umano

8 hairs – llama
Lamahaar (n)
pelo (m) de llama
laine (f) de lama
pelo (m) di lama

9 hairs – mink
Nerz (m)
pelo (m) de visón
poil (m) de vison
pelo (m) di visone

10 hairs – mohair
Mohair (m)
pelo (m) de cabra de angora
mohair (m)
pelo (m) di mohair

11 hairs – mohair, scoured
Haare (n pl) – Mohair gewaschen
pelo (m) descrudado de angora
poils (m pl) – mohair lavé
pelo (m) di mohair lavato

12 hairs – rabbit
Kaninchenhaar (n)
pelo (m pl) de conejo
poil (m) de lapin
pelo (m) di coniglio

13 hairs – vicuna
Vikunahaar (n)
pelo (m pl) de vicuña
poil (m) de vigogne
pelo (m) di vigogna

14 hairs – yak
Yakhaar (n)
pelo (m pl) de yak
poil (m) de yak
pelo (m) di yak

15 hairs – yak dehaired
Haare (n pl) – Yak enthaart
pelo (m pl) de yak pelado
poils (m pl) – yak dépilé
pelo (m) di yak depilato

16 handknitting
Handstricken (n)
punto (m) hecho a mano
tricotage (m) à la main
maglieria (f) a mano

17 handloom
Handwebstuhl (m)
telar (m) manual
métier (m) à bras
telaio (m) a mano

18 handweaving
Handweben (n)
tejer (m) a mano
tissage (m) à la main
tessitura (f) a mano

1 hank
Strähne (f), Strang (m)
madeja (f)
écheveau (m)
matassa (f)

2 hank dyeing
Strähnenfärben (n), Strangfärben (m)
teñidura (f) de madejas
tinture (f) d'echeveaux
tintura (f) in matassa

3 hardness
Härte (f)
dureza (f)
dureté
durezza (f)

4 harness
Harnisch (m)
arreos (m pl)
harnais (m)
rimessa (f) licciata (f)

5 heald
Weblitze (f)
lizo (m)
remisse (m)
liccio (m)

6 health and safety equipment
Gesundheits- und Sicherheitszubehör (n)
equipos (m pl) para la salud y seguridad
équipement (m) de protection sanitaire et de sécurité
apparecchiature (f) per la salute e la sicurezza

7 heat
Hitze (f)
calor (m)
chaleur (f)
calore (m)

8 heat recovery
Warmerückgewinnung (f)
recuperación (f) por calor
récupération (f) de la chaleur
recupero (m) del calore

9 heat setting
Heißfixierung (f)
termofijación (f)
thermofixage (m)
regolazione (f) del calore

10 hemp
Hanf (m)
cáñamo (m)
chanvre (m)
canapa (m)

11 hide
Fell (n)
pellejo (m)
peau (f) de bête
pelle (f) d'animale, cuoio (m)

12 hides
Felle (n pl)
pieles
peaux (f pl) de bêtes
pelli (f pl) d'animali

13 honeycomb weave
Waffelbindung (f)
panal (m)
dessin (m) gaufré
armatura (f) a nido d'ape

14 hopper
Einfülltrichter (m)
tolva (f)
chargeuse (f)
imbuto (m) di alimentazione/ramaggia (f)

15 hose
Strumpf (m); Schlauch (m)
calceta (f); manguera (f)
bas (m); tuyau (m) flexible
calza (f); tubo flessible

16 hosiery
Strumpfwaren (f pl)
calcetería (f)
bonneterie (f)
calzetteria

17 hosiery and knitted goods manufacturers
Strumpf- und Strickwarenhersteller (m pl)
fabricantes (m pl) de calcetería y prendas hechas de punto
fabricants (m pl) de bonneterie et de tricot
produttori (m pl) di calzetteria e prodotti lavorati a maglia

1 hot air
Heißluft (f)
aire (m) caliente
air (m) chaud
aria (f) calda

2 hotels
Hotels (n pl)
hoteles (m pl)
hôtels (m pl)
hotel (m pl)

3 hot water
Heißwasser (n)
agua (f) caliente
eau (f) chaude
acqua (f) calda

4 humid
feucht (adj)
húmedo
humide
umido

5 humidification
Befeuchtung (f)
humidificación (f)
humidification (f)
umidificazione (f)

6 hydraulic
hydraulish (adj)
hidráulico
hydraulique
idraulico

7 hydro-extractor
Schleuder (f), Zentrifuge (f)
hidroextractor (m)
hydro-extracteur (m)
estrattore (m) idraulico

I

8 immersion
Tauchen (n)
inmersión (f)
immersion (f)
immersione (f)

9 import
Import (m)
importación (f)
importation (f)
importazione (f)

10 import clearances
Importdeklaration (f)
despacho (m) de aduana
formalités (f pl) douanières à l'importation
sdoganamento (m) d'importazione

11 importer
Importeur (m)
importador (m)
importateur (m)
importatore (m)

12 industrial photographers
Industriefotografen (m pl)
fotógrafos (m pl) industriales
photographes (m pl) industriels
fotografi (m pl) industriali

13 industrial ventilation
industrielle Ventilation (f) (fig. gewerbliche Erörterung (f)
ventilación (f) industrial
ventilation (f) industrielle
ventilazione (f) industriale

14 inflammable
feuergefährlich, brennbar (adj)
inflamable
inflammable
infiammabile

15 inspection
Prüfung (f)
revisión (f)
contrôle (m)
ispezione (f)

16 instruments – fibre, yarn and fabric testing
Instrumente (n pl) für Faser-, Garn- und Stofftests
instrumentos (m pl) para probar fibras, hilos y telas
instruments (m pl) d'essai des fibres, fils et tissus
strumenti (m pl) per il controllo di fibre, filati e tessuto

17 insulation
Isolierung (f)
aislador (m)
isolation (f)
isolamento (m), coibentazione (f)

1 insurance brokers
Versicherungsvertreter (m pl)
agents (m pl) de seguros
courtiers (m pl) d'assurance
mediatori (m pl) di assicurazione

2 interpretation services
Dolmetscherdienste (m pl)
servicios (m pl) de intérpretes
services (m pl) d'interprétation
servizi (m pl) d'interpretazione

3 investment
Investition (f)
inversión (f)
investissement (m)
investimento (m)

4 invoice
Faktur(a) (f), Rechnung (f)
factura (f)
facture (f)
fattura (f)

J

5 jacquard
Jacquard (m)
jacquard (f)
jacquard (m)
jacquard (m)

6 jacquard fabrics
Jacquardgewebe (n pl)
telas (f pl) jacquard
jacquardées (f pl)
tessuti (m pl) jacquard

7 jacquard loom
Jacquardstuhl (m)
telar (m) jacquard
métier (m) jacquard
telaio (m) jacquard

8 jacquard weave
Jacquardbindung (f)
tejido (m) jacquard
armure (f) jacquard
tessitura (f) jacquard

9 janitorial supplies
Zulieferer an Hausmeister
suministros (m pl) para porteros
fournitures (f pl) auxiliaires
fornitore (f pl) da portineria

10 jersey fabrics
Jerseygewebe (n pl)
telas (f pl) de estambre fino
tricots (m pl) jersey
tessuti (m pl) jersey

11 jet
Jet (m), Strahl (m), Düse (f)
caño (m) de salida
jet (m)
getto (m)

12 jet loom
Jetwebstuhl (m), Düsenwebstuhl (m)
telar (m) de vapor
métier (m) à injection
telaio (m) a getto d'arici

13 jute
Jute (f)
cáñamo (m), yute (m)
jute (m)
iuta (f)

14 jute bags
Jutesäcke (f pl)
sacos (m pl) de cáñamo
sacs (m pl) de jute
iuta (f) – sacchi

15 jute pullers
Juteverzieher (m pl)
tiradores (m pl) de cáñamo
arracheurs (m pl) de jute
iuta (f) – tiratori

16 jute spinning
jutespinnerei (f)
hilatura de cáñamo
filature (f) de jute
iuta (f) – filatura

K

1 kemp
Stichelhaar (n)
lana (f) burda
jarre (m)
fibra (f) ruvida

2 khaki
khaki (adj)
caqui (m)
kaki (m)
kaki (agg)

3 knit (v)
stricken
hacer punto
tricoter
lavorare (m) a maglia, maglia (f), indumento (m) a maglie, maglia (f) diritta

4 knitted fabric
Wirkware (f) – Freizeitkleidung (f)
tejido de punto
tissu (m) tricoté pour vêtements de loisir
tessuto (m) lavorato a maglia – indumenti per il tempo libero

5 knitted fabric – sportswear
Wirkware (f) – Sportkleidung (f)
tejido de punto prendas deportivas
tricots (m pl) de sport
tessuto (m) lavorato a maglia – indumenti sportivi

6 knitted nylon pocketing
geknüpftes Taschenfutter (n) aus Nylon
bolsillos de punto de Nylon
doublures (f pl) pour poches en nylon tricoté
tasche (f pl) di nailon lavorato a maglia

7 knitting machine
Wirk-Nylon (n)
máquina (f) de tricotar
métier (m) à tricoter
macchina (f) per maglieria

8 knitwear manufacturers
Strickwarenhersteller (m pl)
fabricantes (m pl) de prendas hechas de punto
fabricants (m pl) d'articles tricotés
produttori (m pl) di maglieria

9 knops – wool
Noppenwolle (f)
bucles-lana
nopes (f pl) de laine
bottoni (m pl) di lana

10 knops – wool synthetics
Nappenwolle/Synthetikgewebe (n)
bucles sintéticos de lana
nopes (f pl) de laine synthétique
bottoni (m pl) di lana sintetica

11 knot
Knoten (m)
nudo (m)
noeud (m)
nodo (m)

12 knotless
knotenlos (adj)
sin nudos
exempt de noeuds
privo di nodi

13 knotting
Knüpfen (m)
anudar (m)
nouage (m)
decorazione (f) a nodi/annodatura (f)

L

14 label
Etikette (m)
etiqueta (f)
étiquette (f)
etichetta (f)

15 labels
Etiketten (m pl)
etiquetas (f pl)
étiquettes (f pl)
etichette (f pl)

1 laboratory and testing equipment
Labor und Prüfausrüstung (f)
equipos (m pl) de laboratorios y de pruebas
appareillage (m) de laboratoire et de contrôle
apparecchiature (f) da laboratorio ed esame/prove

2 lambswool
Lammwolle (f)
lana (f) de cordero
laine (f) d'agneau
lambswool, lana d'agnello

3 laminators
Kaschiermaschinen (f pl)
laminadores (m pl)
lamineurs (m pl)
laminatori (m pl)

4 leather
Leder (n)
cuero (m)
cuir (m)
pelle (f)

5 leather auxiliaries
Ledernebenprodukte (n pl)
auxiliares (m pl) de cuero
auxiliaires (m pl) en cuir
ausiliari (m pl) di pelle

6 length
Länge (f)
longitud (f)
longueur (f)
lunghezza (f)

7 let off [weaving]
Abzug (m)
suelta (f) de urdimbre
réglage (m) de tension de chaine
regolazione (f) della tensione della catena

8 life insurance and pension consultants
Lebensversicherungs- und Pensionsberater
consultantes (m pl) de seguros y pensiones vitalicias
conseillers (m pl) en assurance sur la vie et la retraite
consulenti (m pl) d'assicurazioni sulla vita e pensioni

9 light
leicht (adj)
ligero
léger
leggero; luce (f)

10 lightweight
leicht; Leichtgewicht (n)
peso (m) ligero
de peu de poids
leggero

11 lining – check
Futter (n) – Karo/Karofutter (n)
forro (m) de cuadros
doublure (f) quadrillée
fodera (f) – a quadretti

12 lining – flame retardant
flammenbeständiges futter (n)
antiinflamable
doublure (f) ignifuge
fodera (f) – antifiamma

13 lining – jacquard
Futter (n) – jacquard, Jacquardfutter (n)
forro (m) jacquard
doublure (f) Jacquard
fodera (f) – jacquard

14 lining – nylon
Nylonfutter (m)
forro (m) nilón
doublure (f) nylon
fodera (f) – nailon

15 lining – polyester
Futter (n) – Polyesterfutter (n)
forro (m) poliéster
doublure (f) polyester
fodera (f) poliestere

16 lining – printed
gedrucktes Futter (n)
forro (m) estampado
doublure (f) imprimée
fodera (f) stampata

17 lining – raised
gerauhtes Futter (n)
forro (m) elevado
doublure (f) grattée
fodera (f) garzata

1 liquid
Flüssigkeit (f)
líquido (m)
liquide (m)
liquido (m)

2 liquid
flüssig (adj)
líquido
liquide
liquido

3 London shrinking
Londoner-Krumpf (m), Krumpfen (n)
encogimiento (m) londinense
décatissage (m) London shrink
restringimento Londinese

4 lubricants
Gleitmittel (n pl), Schmiermittel (n pl)
lubricantes (m pl)
lubrifiants (m pl)
lubrificanti (m pl)

5 lubricants – blending
Schmiermittel (n pl) – Mischung (f)
lubricantes (m pl) de mezcla
lubrifiants (m pl) de mélange
lubrificanti (m pl) – mischia

6 lubricants – dry powder
Schmiermittel (n pl) – Trockenpulver (n)
lubricantes (m pl) tipo polvo seco
lubrifiants (m pl) – poudre inerte
lubrificanti (m pl) – in polvere

7 lubricants – greases and oils
Schmiermittel (n pl) – Fette (n pl) und Öle (n pl)
lubricantes (m pl) – grasas y aceites
lubrifiants (m pl) – graisses et huiles
lubrificanti (m pl) – grassi ed oli

8 lubricants – size
Schmiermittel (n pl) – Leim-Schlichte
lubricantes (m pl) – cola
lubrifiants (m pl) – produits d'encollage
lubrificanti (m pl) – colla, bozzina (f)

9 lubricants – textiles
Schmiermittel (n pl) für Textilien
lubricantes (m pl) textiles
lubrifiants (m pl) textiles
lubrificanti (m pl) – tessili

10 lubricants – yarn
Schmiermittel (n pl) für Garn
lubricantes (m pl) de hilos
lubrifiants (m pl) pour fils
lubrificanti (m pl) – filato

M

11 machine
Maschine (f)
máquina (f)
machine (f)
macchina (f)

12 machinery – accessories
Maschinen (f pl)/Apparaturen (f pl)/ Vorrichtungen (f pl) – Zusatzteile (n pl)
accesorios (m pl) maquinaria
équipement (m) en matériel – accessoires (m pl)
macchinari (m) – accessori

13 machinery – agents
Maschinen (f pl) – Mittel (n pl)
agentes (m pl) de maquinaria
équipement (m) en matériel – agents (m pl)
macchinari (m) – agenti

14 machinery – air splicing
Maschinen (f pl) – für Luft-Spleißen (m)
maquinaria (f) neumática de empalmar
équipement (m) en matériel – renforcement (m) par air
macchinari (m) – giuntura ad aria

15 machinery – aprons
Maschinen (f pl) – Zufuhrtücher (n pl)
maquinaria (f) para fabricar delantales
équipement (m) en matériel – tabliers (m pl)
macchinari (m) – nastri trasportatori a piastre

1 machinery – area weight measurement
Maschinerie (f) für Flächengewichtsmessung
maquinaria (f) para calcular el peso por áreas
matériel (m) – mesure surface poids
macchinari (m) – misurazione del peso di un'area

2 machinery – automatic warp drawing in
Maschinen (f pl) für automatische Ketteneinzieh-vorrichtung
maquinaria (f) automáticamente para tirar urdimbre
équipement (m) en matériel à rentrer les chaînes
macchinari (m) – rimettaggio automatico dell'ordito

3 machinery – bale opening
Maschinen (f pl) für Ballenöffnen
maquinaria (f) para abrir balas
équipement (m) en matériel pour ouvrir les balles
macchinario – apertura delle balle

4 machinery – bale presses
Maschinen (f pl) für Ballenpressen
maquinaria (f) para prensar balas
équipement (m) en matériel – presses (f pl) à emballer
macchinari (m) – presse per balle

5 machinery – balling
Maschinen (f pl) für Knäuelwickler
maquinaria (f) para hacer ovillos
équipement (m) en matériel – pelotonneuses
macchinari (m) – aggomito-latura

6 machinery – batching devices
Maschinen (f pl) für Batschvorrichtung
maquinaria (f) de aparatos de engrasado
équipement (m) en matériel – dispositifs (m pl) à enrouler
macchinari (m) – dispositivi per l'oliaggio

7 machinery – batching motions
Maschinen (f pl) für Batsch-bewegungen
maquinaria (f) de engrase contínuo
équipement (m) en matériel – mouvements (m pl) d'enroulement
macchinari (m) – movimenti di oliaggio

8 machinery – beam back
Maschinen (f pl) Baumrückseite
maquinaria (f) plegador trasero
équipement (m) en matériel – rouleaux (m pl) à ourdir
macchinari (m) – parte posteriore del subbio

9 machinery – beam, beam flanges, jacks, racks, etc.
Maschinen (f pl) für Baum, Baumscheiben, Heber, Streckrahmen
maquinaria (f) plegador, platos del plegador, gatos, cremalleras, etc.
équipement (m) en matériel – ensouple (m), disques (m pl) de l'ensouple, giettes (f pl), crémaillères (f pl), etc.
macchinari (m) – subbio, dischi del subbio, leve degli alberi dei licci, cremagliere, ecc.

10 machinery – beam handling equipment
Gerätschaften für das Bäumen
maquinaria (f) – equipo para manejar el plegador
équipement (m) en matériel – équipement (m) de manutention de l'ensouple
macchinari (m) – attrezzatura di maneggio del subbio

11 machinery – beam – high speed warping
Maschinen (f pl) für Hoch-geschwindigkeitsschärbaum
maquinaria (f) plegador-urdimbre rápida
équipement (m) en matériel – ourdissoirs (m pl) rapides
macchinari (m) – subbio – orditura rapida

1 machinery – beam knitting
Maschinerie (f) für Baumknüpfen
maquinaria (f) – tricotado en el plegador
matériel (m) – tricotage ensouple
macchinari (m) – maglieria in subbio

2 machinery – beam – loom
Maschinen (f pl) für Webbaum
maquinaria (f) plegador-telar
équipement (m) en matériel – ensouple (m) de derrière
macchinari (m) – subbio – telaio

3 machinery – beaming
Maschinen (f pl) für Bäumen
maquinaria (f) de vigas
équipement (m) en matériel d'ensouplage
macchinari (m) – insubbiamento

4 machinery – beams
Maschinen (f pl) für Webbäume
maquinaria (f) sopladores
équipement (m) en matériel – ensouples (m pl)
macchinari (m) – subbi

5 machinery – belting
Maschinen (f pl) – Treibriemen (m)
maquinaria (f) correas
équipement (m) en matériel – courroies (f pl)
macchinari (m) – nastro trasportatorie/materiale (m) per cinghiette

6 machinery – belting drives and V-belt drives
Maschinen (f pl) – Riemenantrieb (m), V-Riemenantrieb (m)
maquinaria (f) de correas de transmisión y de correas tipo 'V'
équipement (m) en matériel – commandes (f pl) à courroie, transmission (f) par courroie trapézoïdale
macchinari (m) – trasmissione a cinghia e trasmissione a cinghia trapezoidale

7 machinery – belting – extensible
dehnbare Treibriemen (m pl) für Maschinen
correas (f pl) extensibles para maquinaria
matériel (m) – courroies extensibles
macchinari (m) – materiale per cinghiette – estensibile

8 machinery – belting – PVC and rubber
Maschinerie (f) für Treibriemen – für PVC und Gummi
correas (f pl) de PVC y goma para maquinaria
matériel (m) – courroies – PVC et caoutchouc
macchinari (m) – materiale per cinghiette – PVC e gomma

9 machinery – blanket
Maschinenmaterial (n) für Decken (m)
maquinaria (f) para mantas
matériel (m) – couvertures
macchinari (m) – coperta

10 machinery – bleaching
Maschinen (f pl) für Bleicher
maquinaria (f) de blanquear
équipement (m) en matériel à blanchir
macchinari (m) – candeggio

11 machinery – blending
Apparaturen (f pl) für Mischerei
maquinaria (f) para la mezcla
équipement (m) en matériel à mélanger
macchinari (m) – mischia

12 machinery – blowroom
Apparaturen (f pl) für Putzerei
maquinaria (f) sala de apertura
équipement (m) en matériel – pour salle de nettoyage
macchinari (m) – sala di soffiaggio

1 machinery – bobbins, cops, etc.
Maschinen (f pl) für Spulen, Kopse usw.
maquinaria (f) de devanaderas, husos, etc.
équipement (m) en matériel – bobines (f pl), canettes (f pl), etc
macchinari (m) – bobine, spole, ecc.

2 machinery – boilers
Maschinen (f pl) für Kessel
maquinaria (f) para calderas
équipement (m) en matériel – chaudières (f pl)
macchinari (m) – caldaie/ bollitori

3 machinery – bolts, nuts, screws, tools, etc.
Maschinerie (f) für Schrauben, Muttern, Bolzen, Werkzeuge usw.
maquinaria (f) para pernos, tuercas, tornillos, herramientas, etc.
matériel (m) – boulons, écrous, vis, outils, etc.
macchinari (m) – bulloni, dadi, viti, utensili, ecc.

4 machinery – braiding
Apparaturen (f pl) für Flechten
maquinaria (f) para trenzar
équipement (m) en matériel à tresser
macchinari (m) – passamaneria

5 machinery – brokers and merchants
Maschinerie (f) für Makler und Händler
maquinaria (f) – agentes, comerciantes
matériel (m) – courtiers et négociants
macchinari (m) – sensali e commercianti

6 machinery – brushing
Maschinen (f pl) Zum Bürsten Und Rauhen
maquinaria (f) para cepillar
équipement (m) en matériel à brosser
macchinari (m) per spazzolatura

7 machinery – bulking
Maschinerie (f) Zum Bauschen
maquinaria (f) para aumentar el volumen
matériel (m) – traitement et voluminosité
macchinari (m) per aumento del volume

8 machinery – calendar bowls
Apparaturen (f pl) für Kalander walten
maquinaria (f) de rodillos de calandrado
équipement (m) en matériel – rouleaux (m pl) de calandre
macchinari (m) – per rulli cilindratori

9 machinery – carbonizing
Apparaturen (f pl) für Karbonisierungsmaschine
maquinaria (f) para carbonizar
équipement (m) en matériel de carbonisation
macchinari (m) – per carbonizzazione

10 machinery – card clothing
Maschinen (f pl) für Kardenbeschlagvorrichtung
maquinaria (f) para ropa de carda
équipement (m) en matériel de montage des garnitures
macchinari (m) – per guarnizione di carda

11 machinery – card- clothing equipment
Maschinen (f pl) für Kardenkontrollausrüstung
maquinaria (f) de equipos para regular la carda
équipement (m) en matériel pour carton de commande
macchinari (m) – apparec- chiatura di guarnizione di carda

1 machinery – card clothing [metallic]
Maschinerie (f) für Krempelbeschlag [metallisch]
maquinaria (f) para telas metálicas de carda
matériel (m) – garniture de carde [métallique]
macchinari (m) – guarnizione di carda [metallica]

2 machinery – card cutters
Maschinen (f pl) für Kartenschläger
maquinaria (f) de cortar la carda
équipement (m) en matériel à piquer les cartons
macchinari (m) – tagliatori di cardo

3 machinery – carding
Apparaturen (f pl) für die Kardiermaschine
maquinaria (f) de cardar
équipement (m) en matériel à carder
macchinari (m) – cardatura

4 machinery – card- and lapping for non-woven carpets
Maschinerie (f) für Kardieren und Lapping für Vliesstoff-Teppiche
maquinaria (f) para cardar y pasar alfombras no tejidas por el baton
matériel (m) – cardeuses et enrouleuses pour tapis non-tissés
macchinari (m) – carda e battitoi per tappeti non tessuti

5 machinery – card control equipment
Maschinerie (f) für Kardierungssteuerungsgeräte
maquinaria (f) para equipos de regular el cardar
matériel (m) – équipement de commande cardes
macchinari (m) – apparecchiatura di controllo delle carde

6 machinery – card grinding fillets
Maschinerie (f) für Kratzenbeschlagschleifabnehmer
maquinaria (f) – filetes para afilar los dientes de la carda
matériel (m) – feuilles à aiguiser les cardes
macchinari (m) – nastri di guarnizione per l'affilatura dei denti delle carde

7 machinery – card grinding rollers
Maschinerie (f) für Kratzenbeschlagschleifrollen
maquinaria (f) para arrolladores de amolar la carda
matériel (m) – cylindres à aiguiser les cardes
macchinari (m) – rulli per l'affilatura dei denti delle carde

8 machinery – card roller repairs
Maschinerie (f) für Kratzenbeschlagrollenreparatur
maquinaria (f) – los rodillos de la carda
matériel (m) – réparation de tambours à dents
macchinari (m) – riparazione dei rulli a carda

9 machinery – cards
Kardiermaschinen (f pl)
maquinaria (f) de cardas
équipement (m) en matériel – cardes (f pl)
macchinari (m) – carde

10 machinery – cards, dobbie, jacquard, etc.
Maschinerie (f) für Kratzen, Schaftmaschinen, Jacquard usw.
maquinaria (f) para cardar tipos dobbie, jacquard, etc.
matériel (m) – cardes Jacquard, cartons de mécanique d'armures, etc.
macchinari (m) – carde, ratiera, jacquard, ecc.

1 machinery – cards, non-woven
Maschinen (f pl) Karden für nicht gewebte stoffe
maquinaria (f) de cardas para los "no-tejidos"
équipement (m) en matériel – cardes (f pl) pour le non-tissé
macchinari (m) – carde per stoffa non tessuta.

2 machinery – cards, semi-worsted
Maschinen (f pl) Halbkammgarn-Karden
maquinaria (f) de cardas medio estambre
équipement (m) en matériel – cardes (f pl) pour le mi-peigné
macchinari (m) – carde per stoffa semi pettinata

3 machinery – cards, single part and tandem
Maschinerie (f) für Kratzen, Einzelteil und Tandem
maquinaria (f) para cardar, sola y tándem
matériel (m) – cardes, simple et tandem
macchinari (m) – carde, parte singola e abbinata

4 machinery – cards, woollen
Wollkarden (f pl) – Maschinen (f pl)
maquinaria (f) de cardas de lana
équipement (m) en matériel – cardes (f pl) pour laine cardée
macchinari (m) – di carde per lana

5 machinery – carpet and rug
Apparaturen (f pl) für Teppich- und Vorlegermaschine
maquinaria (f) de alfombras y alfombrillas
équipement (m) en matériel pour moquettes et tapis
macchinari (m) – tappeti e tappetini

6 machinery – carts, giant batch
Maschinerie (f) für Kratzen, Riesenchargen
carros (m pl) para transportar lotes grandes
matériel (m) – chariots pour lots géants
macchinari (m) – carrelli per il trasporto di ingenti quantitativi

7 machinery – cleaners, travelling overhead
Maschinerie (f) für Reiniger, Overhead-Betrieb
limpiadores (m pl) de maquinaria que pasan por encima
matériel (m) – nettoyeuses roulantes suspendues
macchinari (m) – battitrici – anellino di testa

8 machinery – cloth baling
Maschinen (f pl) für Stoff – Ballenpresse
maquinaria (f) de embalar tela
équipement (m) en matériel de mise en balles du tissu
macchinari (m) – imballaggio del tessuto

9 machinery – cloth batching
Maschinen (f pl) für Stoff – Batschvorrichtung
maquinaria (f) de agrupación de tela
équipement (m) en matériel d'enroulage du tissu
macchinari (m) – oliaggio del tessuto

10 machinery – cloth beaming
Maschinerie (f) für Stoffbäumen
maquinaria (f) para plegar tela
matériel (m) – ensouplage du tissu
macchinari (m) – in subbiamento di Staffa

11 machinery – cloth blowing
Maschinen (f pl) für Stoff-Reiniger
maquinaria (f) para vaporar las telas
équipement (m) en matériel à souffler les tissus
macchinari (m) – soffiaggio del tessuto/stoffa

12 machinery – cloth brushing
Tuchbürsten (f pl) Gerätschaften (f pl)
maquinaria (f) para cepillar tela
équipement (m) en matériel à brosser les poils
macchinari (m) – spazzolatura del tessuto

1 machinery – cloth crabbing
Apparaturen (f pl) für Tuch-Krappen
maquinaria (f) para arrollar tela
équipement (m) en matériel de lissage
macchinari (m) – crabbing del tessuto/or-fissaggio a umido

2 machinery – cloth cutting
Apparaturen (f pl) als Tuch Zuschneide-Maschine
maquinaria (f) de cortar tela
équipement (m) en matériel à découper les tissus
macchinari (m) – taglio del tessuto

3 machinery – cloth damping
Tuchbefeuchtungs-Maschinen (f pl)
maquinaria (f) para humedecer tela
équipement (m) en matériel à mouiller les tissus
macchinari (m) – inumidimento del tessuto/umidificazione del tessuto

4 machinery – cloth drying
Tuchtrockner – Maschines (f pl)
maquinaria (f) para secar tela
équipement (m) en matériel à sécher les tissus
macchinari (m) – asciugamento del tessuto

5 machinery – cloth dyeing
TuchfFärber-Maschinen (f pl)
maquinaria (f) de teñir tela
équipement (m) en matériel de teinture de tissu
macchinari (m) – tintura del tessuto

6 machinery – cloth examining
Tuchprüf-Maschinerie (f)
maquinaria (f) para examinar telas
équipement (m) en matériel de vérification des tissus
macchinari (m) – esame del tessuto

7 machinery – cloth finishing
Maschinen (f pl) für Tuch-Erdausrüstung
maquinaria (f) para el acabado de telas
équipement (m) en matériel de finissage de tissu
macchinari (m) – finissaggio del tessuto

8 machinery – cloth fulling
Apparaturen (f pl) für Tuch-Walken
maquinaria (f) para abatanar las telas
équipement (m) en matériel de foulage du drap
macchinari (m) – follatura del tessuto

9 machinery – cloth guiding
Maschinen (f pl) mit Tuch-Führungsschiene
maquinaria (f) para guiar tela
équipement (m) en matériel – guide-tissu (m)
macchinari (m) – guida/ passaggio del tessuto

10 machinery – cloth hydro extracting
Tuchschleuder (f) – Maschinen (f pl)
maquinaria (f) para la hidroextracción de tela
équipement (m) en matériel à centrifuger
macchinari (m) – idroestrazione del tessuto

11 machinery – cloth laying, measuring and rolling
Tuchlege, Meß- und Rollenanlage (f) – Maschinen (f pl)
maquinaria (f) para tender, medir y arrollar la tela
équipement (m) en matériel à poser, mesurer et enrouler les tissus
macchinari (m) – avvolgimento, misurazione, arrotolamento del tessuto

1 machinery – cloth milling
Maschinen (f pl) – Tuch-Walkmaschine (f)
maquinaria (f) para el batanado del tejido
équipement (m) en matériel de foulage du drap
macchinari (m) – follatura del tessuto

2 machinery – cloth monitoring systems
Maschinen (f pl)) für Tuch-Überwachungssysteme
maquinaria (f) sistemas para controlar el tejido
équipement (m) en matériel – systèmes (m pl) de contrôle du tissu
macchinari (m) – sistemi di monitorizzazione del tessuto

3 machinery – cloth presses – paper-press
Maschinerie (f) für Stoffpressen – Papierpresse
maquinaria (f) para prensar tela y papel
matériel (m) – presses à tissu – presses à papier
macchinari (m) – presse per stoffe e per carta

4 machinery – cloth presses – rotary
Maschinerie (f) – für Stoffpressen – Karussell
maquinaria (f) rotativa para prensar tela
matériel (m) – presses à tissu rotatives
macchinari (m) – presse per stoffe – rotante

5 machinery – cloth rigging and rolling
Maschinen (f pl) – Tuch- und Rollvorrichtungs-Maschinen (f pl)
maquinaria (f) para arreglar y arrollar tela (f)
équipement (m) en matériel à plier et à enrouler les tissus
macchinari (m) – faldatura ed arrotolamento del tessuto

6 machinery – cloth rolling and lapping
Apparaturen (f pl) für Tuch- Roll- und Wickelmaschine
maquinaria (f) para arrollar y traslapar tela
équipement (m) en matériel à enrouler et à noper les tissus
macchinari (m) – arrotolamento e smerigliatura del tessuto

7 machinery – cloth roll packing
Maschinerie (f) für Stoffrollenverpackung
maquinaria (f) para embalar rollos de tela
matériel (m) – emballage de rouleaux de tissu
macchinari (m) – imballaggio dei rotoli di stoffa

8 machinery – cloth scouring
Apparaturen (f pl) für Tuch-Reinigungsmaschine
maquinaria (f) para descrudar tela
équipement (m) en matériel de débouillissage des tissus
macchinari (m) – lavaggio del tessuto

9 machinery – cloth slitting, open-width and tubular
Maschinen (f pl) für Tuch-Streifenschneider (m), offene Breite, schlauchförmig
maquinaria (f) para hender tela estirada o tubular
équipement (m) en matériel à refendre les tissus au large et tubulaire
macchinari (m) – tagliare a nastri il tessuto, tessuto aperto e tubolare

10 machinery – cloth tentering
Tuchspannvorrichtungs-Maschinen (f pl)
maquinaria (f) para estirar tela
équipement (m) en matériel – élargisseuses (f pl)
macchinari (m) – asciugatura del tessuto

1 machinery – coilers
Wickler (m pl) – Maschinen (f pl)
maquinaria (f) para arrollar
équipement (m) en matériel – pots (m pl) tournants
macchinari (m) – bobinatrici

2 machinery – combing
Kämmvorrichtungs-Maschinen (f pl)
maquinaria (f) para cardar
équipement (m) en matériel à peigner
macchinari (m) – pettinatura

3 machinery – combing leathers
Maschinerie (f) für Lederkämmen
maquinaria (f) para peinar piel
matériel (m) – manchons pour peigneuses
macchinari (m) – pettinatura di pelle

4 machinery – combs
Maschinen (f pl) für Kämme
maquinaria (f) para cardar
équipement (m) en matériel – peignes (f pl)
macchinari (m) – pettini

5 machinery – condenser bobbins
Vorgarnwickel-Maschinen (f pl)
maquinaria (f) para bobinas de condensar
équipement (m) en matériel – bobines (f pl) de condenseur
macchinari (m) – bobine condensatori

6 machinery – condenser rubbers
Maschinen (f pl) für Vorspinngummi
maquinaria (f) de gomas de condensadores
équipement (m) en matériel – frotteurs (m pl) de condenseur
macchinari (m) – gomme per carde finitrici

7 machinery – condenser tapes
Maschinen (f pl) für Kondensertiemchen
maquinaria (f) de cintas de condensadores
équipement (m) en matériel – lanières (f pl) de carde fileuse
macchinari (m) – nastri per carde finitrici

8 machinery – condensers
Maschinen (f pl) für Kondensatoren
maquinaria (f) de condensadores
équipement (m) en matériel – condenseurs (m pl)
macchinari (m) – condensatori

9 machinery – cone and pirn winding
Maschinen (f pl) für Konen- und Schloßspuleinrichtung
maquinaria (f) para devanar carretes y husos
équipement (m) en matériel – bobinoirs (m pl) à fil croisé et canetières (f pl)
macchinari (m) – confezione di rocchetti e cannette

10 machinery – cones
Maschinen (f pl) für Kegel, Konen
maquinaria (f) – conos
équipement (m) en matériel – cônes (m pl)
macchinari (m) – rocchetti

11 machinery – continuous washing
Dauerwasch-Maschinen (f pl)
maquinaria (f) para lavado continuo
équipement (m) en matériel – lavage (m) à la continue
macchinari (m) – lavaggio continuo

12 machinery – control systems
Kontrollsystem-Maschinen (f pl)
maquinaria (f) de sistemas de regulación
équipement (m) en matériel – systèmes (m pl) de commande
macchinari (m) – sistemi di controllo

1 machinery – conveying mechanical
mechanische Förderanlange (f)
maquinaria (f) de transporte mecánico
équipement (m) en matériel – convoyeurs (m pl) mécaniques
macchinari (m) – convoglitore meccanico

2 machinery – conveying pneumatic
pneumatische Förderanlage (f)
maquinaria (f) de transporte neumático
équipement (m) en matériel – convoyeurs (m pl) pneumatiques
macchinari (m) – convogliatore pneumatics

3 machinery – conveyor belting
Förderbandanlage – Maschinen (f pl)
maquinaria (f) de correas de transmisión
équipement (m) en matériel – bandes (f pl) de convoyeur
macchinari (m) – materiale per or cinghie nastri trasportatori

4 machinery – conveyor lattice
Förderbandgitter (n) – Maschinen (f pl)
maquinaria (f) para la telera de transmisión
équipement (m) en matériel – tablier (m) transporteur
macchinari (m) – graticcio per nastri trasportatori

5 machinery – cot and apron
Gestell und Zuführtuch (n) – Vorrichtung (f)
maquinaria (f) para manguitos y bolsas de estirar
équipement (m) en matériel – garnitures (f pl) de rouleau et tabliers (m pl)
macchinari (m) – manicotto e cinghietta (f)

6 machinery – counting instruments
Maschinen (f pl) für Zählvorrichtungen
maquinaria (f) de calculadoras
équipement (m) en matériel – dispositifs à compter
macchinari (m) – strumenti contatore

7 machinery – crane hire
Maschinerie (f) in der Kranvermietung
grúas (f pl) de alquiler
matériel (m) – location des grues
macchinari (m) – affitto gru

8 machinery – creels
Maschinen (f pl) für Aufsteckgatter
maquinaria (f) filetas
équipement (m) en matériel – râteliers (m pl) porte-ensouple
macchinari (m) – cantre

9 machinery – crochet
Apparaturen (f pl) für Häkelmaschinen
maquinaria (f) ganchillo
équipement (m) en matériel à crocheter
macchinari (m) – uncinetto

10 machinery – custom built
Apparaturen (f pl) für nach Maß angefertigte Maschinen
maquinaria (f) hecha sobre pedido
équipement (m) en matériel fait sur demande
macchinari (m) – prodotto su richiesta

11 machinery – cutters longitudinal and transverse
Maschinerie (f) für Längs- und Querschneider
talladores (m pl) longitudinales y transversales
matériel (m) – coupeuses longitudinales et transversales
macchinari (m) – frese longitudinali e trasversali

12 machinery – decatising
Apparaturen (f pl) für Dekatiermaschine
maquinaria (f) de prensa a vapor
équipement (m) en matériel à décatir
macchinari (m) – decatissaggio

1 machinery – direct cabling
Maschinerie (f) für Direktverkabelung
maquinaria (f) para el torcido directo
matériel (m) – câblage direct
macchinari (m) – ritorcitura diretta

2 machinery – dismantling and relocation
Maschinen (f pl) für Abbau und Versetzen
maquinaria (f) de desmontar y recolocar
équipement (m) en matériel – démontage (m) et remise (f) en place
macchinari (m) – smontaggio e rilocazione

3 machinery – dobbies, lags, etc.
Apparaturen (f pl) für Schaftmaschinen, Verkleider usw.
maquinaria (f) de aparatos dobby, retardadores, etc.
équipement (m) en matériel – ratières (f pl) d'armures, planchettes (f pl) de garnissage, etc.
macchinari (m) – ratiere, rivestimento dei cilindri, ecc doge (f pl)

4 machinery – doubling
Apparaturen (f pl) für Doubliermaschine
maquinaria (f) para el doblaje
équipement (m) en matériel de doublage
macchinari (m) – accoppiamento (m)

5 machinery – drafting aprons
Maschinen (f pl) für Zuführtücher
maquinaria (f) – bolas de estiraje
équipement (m) en matériel – manchons (m pl) d'étirage
macchinari (m) – nastri trasportatori per stiratori

6 machinery – drawframes
Maschinen (f pl) für das Streckwerk
maquinaria (f) de estirar
équipement (m) en matériel – bancs (m pl) d'étirage
macchinari (m) – stiratoi

7 machinery – drawing
Maschinen (f pl) für Strecken
maquinaria (f) de estiraje
équipement (m) en matériel d'étirage
macchinari (m) – stiro

8 machinery – drawing in
Maschinen (f pl) für Fadeneinzugsvorrichtung
maquinaria (f) de remetido
équipement (m) en matériel à rentrer les chaînes
macchinari (m) – rimettaggio

9 machinery – drop wires
Fadenreiter (m) – verrichtung
maquinaria (f) – laminillas
équipement (m) en matériel – lamelles (f pl) casse-chaîne
macchinari (m) – rimettaggio

10 machinery – drying
Trockner (m) – Maschinen (f pl)
maquinaria (f) de secar
équipement (m) en matériel à sécher
macchinari (m) – essiccamento (m)

11 machinery – drying cylinders
Maschinen (f pl) Trocknerzylinder (m pl) Maschinen (f pl)
maquinaria (f) – cilindros de secado
équipement (m) en matériel – tambours (m pl) de séchage
macchinari (m) – cilindri essiccatori

12 machinery – drying ovens
Maschinerie (f) – Trockenöfen (m pl)
hornos (m pl) de secar
matériel (m) – cuves de séchage
macchinari (m) – forni di essiccamento (m)

1 machinery – dyeing
Färbemaschinen (f pl)
maquinaria (f) para teñir
équipement (m) en matériel pour la teinturerie
macchinari (m) – tintura

2 machinery – dyeing laboratory
Maschinen (f pl) für Farblabor
maquinaria (f) para el laboratorio de tinte
équipement (m) en matériel pour le laboratoire de teinturerie
macchinari (m) – laboratorio di tintura

3 machinery – dyeing ribbon
Maschinerie (f) für Färbeband
maquinaria (f) – cintas para el tinte
matériel (m) – rubans de teinture
macchinari (m) – nastro colorante

4 machinery – expander rollers
Maschinerie (f) für Spannrollen
maquinaria (f) rodillos para el alisado
matériel (m) – rouleaux élargisseurs
macchinari (m) – rulli allargatori

5 machinery – exporters
Apparaturen (f pl) für Maschinenexporteur
maquinaria (f) para exportadores
équipement (m) en matériel – exportateurs (m pl)
macchinari (m) – esportatori

6 machinery – extrusion
Maschinen für Strangpressern, Extrusions-Maschinen (f pl)
maquinaria (f) extrusión
équipement (m) en matériel d'extrusion
macchinari (m) – estrusione

7 machinery – fabric inspection
Maschinen (f pl) für Stoffprüfeinrichtung
maquinaria (f) para revisar telas
équipement (m) en matériel de vérification du tissu
macchinari (m) – ispezione del tessuto

8 machinery – fabricators
Maschinen (f pl) für Hersteller
fabricantes (m pl) de maquinaria
équipement (m) en matériel – fabricants (m pl)
macchinari (m) – fabbricatori

9 machinery – fallers
Maschinen (f pl) für Nadelstäbe
maquinaria (f) para enderezar las fibras en la carda
équipement (m) en matériel – barres (f pl) à aiguilles + barrettes (f)
macchinari (m) – bacchette trasversali/tenditori a pettine

10 machinery – fancy twisters
Maschinen (f pl) für Effektzwirner
maquinaria (f) de retorcer diseños especiales
équipement (m) en matériel à retordre le fil fantaisie
macchinari (m) – ritorcitoi per ritorti fantasia

11 machinery – fancy yarns
Maschinen (f pl) für Ziergarne Effektgarne (n pl)
maquinaria (f) de hilos de fantasía
équipement (m) en matériel – fil (m) fantaisie
macchinari (m) – filati fantasia

12 machinery – fans
Maschinen (f pl) für Fächer /Ventilatoren
maquinaria (f) de ventiladores
équipement (m) en matériel – ventilateurs (m pl)
macchinari (m) – ventilatori

13 machinery – fans – dust extractors
Maschinerie (f) für Gebläse – Staubextraktoren
ventiladores (m pl) y extractores (m pl) de polvo
maquinaria (f) – ventilateurs – dépoussiéreurs
macchinari (m) – aspiratori della polvere

1 machinery – fans, wool conveying
Maschinen (f pl) für Woll-Ventilatortransport
maquinaria (f) de ventiladores para el transporte de lana
équipement (m) en matériel – soufflantes (f pl) pour le transport de la laine
macchinari (m) – ventilatori per il trasporto della lana

2 machinery – fearnought
Krempelwolf (m) – Maschinen (f pl)
maquinaria (f) Bamdor cardador
équipement (m) en matériel – loup-carde (m)
macchinari (m) – sfibratrice per lana

3 machinery – feeding
Maschinen (f pl) – Zuführvorrichtung (f)
maquinaria (f) de alimentación
équipement (m) en matériel d'avance, d'alimentation
macchinari (m) – alimentazione

4 machinery – fibre cutting
Faserschneider (m), Maschinen (f pl)
maquinaria (f) de cortar fibras
équipement (m) en matériel à découper les fibres
macchinari (m) – taglio delle fibre

5 machinery – fibre hydro extracting
Maschinen (f pl) für Faserhydroschleuder
maquinaria (f) de hidroextracción de fibras
équipement (m) en matériel à centrifuger les fibres
macchinari (m) – idroestrazione delle fibre

6 machinery – fibre oiling
Maschinen (f pl) für Faserfettung
maquinaria (f) para lubrificar fibras
équipement (m) en matériel d'ensimage (m) des fibres
macchinari (m) – oliatura delle fibre

7 machinery – fibre reclaiming
Maschinen (f pl) für Faserwiedergewinnung
maquinaria (f) para la recuperación de fibras
équipement (m) en matériel de récupération des fibres
macchinari (m) – ricupero delle fibre

8 machinery – flat and rotary screen printing
Maschinerie (f) für Flach- und Drehsiebdruck
maquinaria (f) para estampar de pantallas planas y rotativas
matériel (m) – sérigraphie à cadre rotatif et à plat
macchinari (m) – stampaggio su schermo piatto e rotante

9 machinery – fleece opening
Maschinen (f pl) für Vliesöffner
maquinaria (f) para abrir el vellón
équipement (m) en matériel – coupeurs (m pl) de nappe
macchinari (m) – apertura del vello

10 machinery – flock and rag
Maschinen (f pl) für Flocker- und Lumpen-Bearbeitung
maquinaria (f) para borra y trapos
équipement (m) en matériel pour le floc (m) et le chiffon (m)
macchinari (m) – fiocco e straccio

11 machinery – friction open-ended spinning
Maschinerie (f) für Friktions-OE-Rotorspinnen
maquinaria (f) de hilar por fricción, extremos abiertos
matériel (m) – filatures à fibres libérées par friction
macchinari (m) – filatura a ingranamento di pezze aperte

1 machinery – fringeing
Apparaturen (f pl) – für Fransendrehmaschine (f)
maquinaria (f) para guarnecer con flecos
équipement (m) en matériel à franger
macchinari (m) – orlatura

2 machinery – fusing
Maschinen (f pl) für Schmelzanlage
maquinaria (f) para combinar
équipement (m) en matériel de fusion
macchinari (m) – spoletta

3 machinery – garnetting
Garnettöffner (m) – Maschinen (f pl)
maquinaria (f) para abrir desperdicios
équipement (m) en matériel – garnetteuses (f pl)
macchinari (m) – garnettatrice

4 machinery – gas conversions
Maschinerie (f) – Gaskonversions -maschinen (f pl)
maquinaria (f) para conversiones de gas
matériel (m) – conversions de gaz
macchinari (m) – conversioni di gas

5 machinery – gilling boxes etc.
Maschinen (f pl) für Nadelstreckenbehandlung
maquinaria (f) de cajas de lana peinada etc.
équipement (m) en matériel – étireuses (f pl) à barrettes, etc.
macchinari (m) – banco di tiratura barrette

6 machinery – gimp and trimming
Maschinerie (f) für Umflechten und Aufputzen
maquinaria (f) para alamar y recortar
matériel (m) – ganse et passementerie
macchinari (m) – filato fantasia e sellatura a finizione

7 machinery – gripper shuttle systems
Maschinerie (f) für Greiferschützensysteme
maquinaria (f) de sistemas de lanzadoras con pinzas
matériel (m) – systèmes de navettes à pince
macchinari (m) – sistemi di navette a pinza

8 machinery – guards
Maschinen (f pl) für Schutzvorrichtungen
rejas (f pl) o chapas (f pl) de seguridad para maquinaria
équipement (m) en matériel – dispositifs (m pl) de protection
macchinari (m) – meccanismi di arresto/schermo protettivo

9 machinery – hair carding opening
Maschinerie (f) für Haarkardierungsöffnung
maquinaria (f) para abrir pelo cardado
matériel (m) – ouverture et cardage des poils
macchinari (m) – apertura di cardatura di pelo

10 machinery – hair trucks
Maschinerie (f) für Haargestelle
carretillas (f pl) para pelos
matériel (m) – chariots pour poils
macchinari (m) – carri trasportatori di peli

11 machinery – hand looms
Maschinen (f pl) für Handwebstühle
telares (m pl) manuales
équipement (m) en matériel – métiers (m pl) à bras
macchinari (m) – telai a mano

12 machinery – harness weaving
Maschinen (f pl) für Harnischweben
maquinaria (f) de telares
équipement (m) en matériel – harnais (m) de tissage
macchinari (m) – tessitura rimessa/licciata (f)

1 machinery – healds and reeds
Maschinen (f pl) für Litzen, Blätter
maquinaria (f) para lizos y peines
équipement (m) en matériel – lisses (f pl) et guide-fils (m pl)
macchinari (m) – licci e pettini

2 machinery – heat setting
Maschinen (f pl) – Thermofixierer (m)
maquinaria (f) termofijación
équipement (m) en matériel de thermofixage
macchinari (m) – fissaggio a caldo

3 machinery – hoists and cranes
Maschinerie (f) – Flaschenzüge und Kranmaschinerie (f)
elevadores (m pl) y grúas (f pl)
matériel (m) – équipement de levage et grues
macchinari (m) – elevatori e gru

4 machinery – hopper feeders
Maschinen (f pl) für Kastenspeiser
maquinaria (f) de alimentación por tolva
équipement (m) en matériel – chargeuses (f pl) à alimentation automatique
macchinari (m) – alimentatori della tramoggia

5 machinery – hosiery
Maschinerie (f) für Strumpf- und Trikotwaren (f pl)
maquinaria (f) para calcetería
matériel (m) – bonneterie
macchinari (m) – calzetteria (f)

6 machinery – humidifiers
Befeuchter (m pl) – Maschinen (f pl)
maquinaria (f) para humedecer
équipement (m) en matériel – humidificateurs (m pl)
macchinari (m) – umidificatori

7 machinery – hydraulics
Hydraulik (f) – Maschinen (f pl)
maquinaria (f) hidráulica
équipement (m) en matériel hydraulique
macchinari (m) – idraulici

8 machinery – hydro extractors
Hydroschleuder (f) – Maschinen (f pl)
maquinaria (f) hidroextractora
équipement (m) en matériel – hydro-extracteurs (m pl)
macchinari (m) – idroestrattori

9 machinery – importers
Apparaturen (f) für Maschinenimporteure
importadores (m pl) de maquinaria
équipement (m) en matériel – importateurs (m pl)
macchinari (m) – importatori

10 machinery – industrial sewing
Maschinen (f pl) für Industrien-ähen
maquinaria (f) de coser industrial
équipement (m) en matériel – machines (f pl) à coudre industrielles
macchinari (m) – cucito industriale

11 machinery – installation
Maschineninstallation (f)
instalación (f) de maquinaria
équipement (m) en matériel – installation (f)
macchinari (m) – installazione

12 machinery – installation engineers
Maschinerie (f) für Installationstechniker
mecánicos (m pl) para montar maquinaria
matériel (m) – installateurs
macchinari (m) – installazione tecniche

13 machinery – insulation
Maschinen (f pl) für Isolierung
maquinaria (f) para aislar
équipement (m) en matériel – isolation
macchinari (m) – isolamento

14 machinery – Jacquard
Apparaturen (f) für Jacquardmaschine
maquinaria (f) jacquard
équipement (m) en matériel – Jacquard
macchinari (m) – jacquard

1 machinery – Jacquard harness
Maschinen (f pl) für Jacquardharnisch
maquinaria (f) de telares jacquard
équipement (m) en matériel – harnais (m) Jacquard
macchinari (m) – jacquard liccio

2 machinery – Jacquarettes
Maschinen (f pl) für Jacquardretten
maquinarias (f) pequeña Jacquard
équipement (m) en matériel – Jacquarettes (f pl)
macchinari (m) – licci jacquard

3 machinery – knitted – circular
Runelstrick (m) – Maschinen (f pl)
maquinaria (f) circular de hacer punto
équipement (m) en matériel – métiers (m pl) circulaires
macchinari (m) – tessuto – circolare maglieria

4 machinery – knitted cloth finishing
Maschinerie (f) für Wirkstoff-Veredlung
maquinaria (f) para rematar paño de punto
matériel (m) – finissage des tissus à maille
macchinari (m) – finissaggio dei tessuti a maglia

5 machinery – knitting
Maschinen (f pl) für Wirkerei
maquinaria (f) de hacer punto
équipement (m) en matériel – métiers (m pl) à tricoter
macchinari (m) – lavorazione a maglia

6 machinery – knotted cloth finishing
Maschinen (f pl) für Knüpfendfertigung
maquinaria (f) – acabado de nudo
équipement (m) en matériel de finissage (m) des tricots
macchinari (m) – finissaggio del tessuto a nodi

7 machinery – laboratory and testing equipment
Maschinen (f pl) als Labor- und Prüfausrüstung
maquinaria (f) para laboratorio y de hacer pruebas
équipement (m) en matériel – équipement (m) de laboratoire et de contrôle
macchinari (m) – attrezzatura per laboratorio e di verifiche/prove

8 machinery – laminators
Apparaturen (f) für Kaschiermaschinen
maquinaria (f) de laminar
équipement (m) en matériel à laminer
macchinari (m) – laminatori

9 machinery – lap layers
Maschinerie (f) für Vliestäfler, Florwickler (m)
maquinaria (f) para tender rollos de fibras
matériel (m) – nappeurs
macchinari (m) – strati di faldi

10 machinery – lattice sheets
Maschinen (f pl) für Lattentücher
maquinaria (f) de láminas de telera
équipement (m) en matériel – panneaux (m pl) de tablier
macchinari (m) – lastre di gratticcio

11 machinery – loom and weaving
Maschinen (f pl) für Web– und Webstuhlanlage
maquinaria (f) telares y de tejer
équipement (m) en matériel pour les métiers et le tissage
macchinari (m) – telaio e tessitura

12 machinery – loom and weaving accessories
Maschinen (f pl) für Web- und Webstuhlzusatzteile
accesorios (m pl) de maquinaria de telares y de tejer
équipement (m) en matériel – accessoires (m pl) pour les métiers et le tissage
macchinari (m) – accessori per il telaio e la tessitura

1 machinery – loom buffers
Maschinerie (f) für Webstuhlpuffer
pulidores (m pl) de telares
matériel (m) – amortisseurs pour métiers à tisser
macchinari (m) – paracolpi del telaio

2 machinery – looms for heavy industrial fabrics
Maschinerie (f) für Webstühle für Schwerweberei
telares (m pl) para telas pesadas e industriales
matériel (m) – métiers pour les tissus lourds industriels
macchinari (m) – telai per tessuti industriali (pesanti)

3 machinery – magnetic separators
Maschinerie (f) für Magnetseparatoren
maquinaria (f) de separadores magnéticos
matériel (m) – séparateurs magnétiques
macchinari (m) – separatori magnetici

4 machinery – mangle bowls
Maschinerie (f) für Mangelwalzen
rodillos (m pl) de calandrar
matériel (m) – rouleaux de mangle
macchinari (m) – ruote-a lanterna

5 machinery – mangles full width
Maschinen (f pl) für Quetschen volle Breite
maquinaria (f) de mangles en su ancho total
équipement (m) en matériel – exprimeuses (f pl) au large
macchinari (m) – mangani – larghezza completa

6 machinery – mangles padding
Maschinen (f pl) für Quetschpolsterung
maquinaria (f) impregnación con mangle
équipement (m) en matériel calandres (m pl) de foulardage
macchinari (m) – imbottitura/ ingrassaggio mangani

7 machinery – mangles squeeze pad
Maschinen (f pl) für Quetschpolster
maquinaria (f) rellenos de exprimir de calandrias
équipement (m) en matériel – foulards-exprimeurs (m pl)
macchinari (m) – tampone di compressione dei mangani

8 machinery – mercerising
Merzerisieranlage
maquinaria (f) de mercerizar
équipement (m) en matériel de mercerisage
macchinari (m) – mercerizza-zione

9 machinery – metal detectors
Metalldetektormaschinerie (f)
maquinaria (f) de detectores de metales
matériel (m) – détecteurs de métaux
macchinari (m) – rivelatori di metalli

10 machinery – milling
Maschinen (f pl) für Walken
maquinaria (f) de desmenuzar
équipement (m) en matériel de foulage
macchinari (m) – feltratura

11 machinery – mixers
Maschinerie (f) für Mixer
maquinaria (f) para mezclar
matériel (m) – mélangeuses
macchinari (m) – agitatori

12 machinery – modernised
modernisierte maschinenaulage (f)
maquinaria (f) modernizada
équipement (m) en matériel modernisé
macchinari (m) – modernizzato

13 machinery – moisture measurement
Maschinen (f pl) für Feuchtigkeitsmessung
maquinaria (f) de medir humedad
équipement (m) en matériel de mesure d'humidité
macchinari (m) – misurazione dell'umidità

1 machinery – monitoring systems
Maschinen (f pl) für Überwachungssysteme
maquinaria (f) de sistemas de control
équipement (m) en matériel – systèmes (m pl) de contrôle
macchinari (m) – sistemi di controllo

2 machinery – mules spinning
Maschinen (f pl) für Selfaktor der Streichgarnspinnerei (f)
maquinaria (f) de hiladores intermitentes
équipement (m) en matériel – renvideurs (m pl) automatiques
macchinari (m) – filatura a filatoio intermittente

3 machinery – napping
Maschinen (f pl) für Rauherei (f)
maquinaria (f) de perchar
équipement (m) en matériel à gratter
macchinari (m) – garzatura

4 machinery – needle loom
Maschinen (f pl) für Nadelwebstuhl
maquinaria (f) de telares de agujas
équipement (m) en matériel – aiguilleteuse (f)
macchinari (m) – telaio ad aghi

5 machinery – needlepoint willeys
Maschinerie (f) für Nadelspitzenreißwölfe
maquinaria (f) para limpiar puntos de agujas
matériel (m) – loups aiguille
macchinari (m) – diavolotti con la punta ad ago

6 machinery – non-automatic
Maschinen (f pl) für nicht automatische Anlage
maquinaria (f) no automática
équipement (m) en matériel non-automatique
macchinari (m) – non automatico

7 machinery – non-automatic loom
Maschinen (f pl) für nicht automatischern Webstuhl
maquinaria (f) de telares no automáticos
équipement (m) en matériel – métiers (m pl) non-automatiques
macchinari (m) – telaio non automatico

8 machinery – non-woven
Maschinen (f pl) für nicht gewebte Stoffe
maquinaria (f) para tela no tejida
équipement (m) en matériel pour le non-tissé (m)
macchinari (m) – stoffa non tessuta

9 machinery – non-woven [dry laid]
Maschinerie (f) für Verbundstoff [auf trockenem Wege]
maquinaria (f) para no tejidos [en seco]
matériel (m) – non-tissé [à sec]
macchinari (m) – non tessuto [posa a seco]

10 machinery – oiling
Apparaturen (f pl) für Schmieren /Schmiermaschine (f)
maquinaria (f) de lubrificar
équipement (m) en matériel d'ensimage
macchinari (m) – oliatura

11 machinery – opening
Öffnervorrichtungsanlage (f)
maquinaria (f) de abrir
équipement (m) en matériel – ouvreuses (f pl)
macchinari (m) – apertura

12 machinery – opening and blending plants
Maschinerie (f) für Öffnungs- und Mischanlagen
maquinaria (f) de abrir y mezclar
matériel (m) – ouvreuses et mélangeuses
macchinari (m) – impianti di apertura e mischie

1 machinery – opening, synthetic-
Maschinerie (f) für Öffnung von Synthetik
maquinaria (f) de abrir sintéticos
matériel (m) – ouvreuses pour synthétique
macchinari (m) – sintetico di apertura

2 machinery – opening, waste-
Maschinerie (f) für Öffnung von Ausschuß
maquinaria (f) de abrir residuos
matériel (m) – effilocheuses pour déchets et fils
macchinari (m) – scarto di apertura

3 machinery – opening, wool-
Maschinerie (f) für Öffnung von Wolle
maquinaria (f) de abrir lana
matériel (m) – ouvreuses pour laine
macchinari (m) – apritoio per la lana

4 machinery – packaging
Verpackungsanlage (f), Maschinen (f pl)
maquinaria (f) de embalar
équipement (m) en matériel d'emballage
macchinari (m) – imballaggio

5 machinery – packing and inspection
Maschinerie (f) für Verpackung und Inspektion
maquinaria (f) de embalar y revisar
matériel (m) – emballage et vérification
macchinari (m) – imballaggio ed ispezione

6 machinery – packing automatic
Maschinerie (f) für automatische Verpackung
maquinaria (f) de embalar automática
matériel (m) – emballage automatique
macchinari (m) – imballaggio automatico

7 machinery – padding
Maschinen (f pl) für Polsterung
maquinaria (f) de rellenos
équipement (m) en matériel de foulardage
macchinari (m) – imbottitura

8 machinery – pattern cutting
Apparaturen (f) für Musterschneidemaschine
maquinaria (f) de cortar diseños
équipement (m) en matériel à découper les patrons
macchinari (m) – taglio dei modelli

9 machinery – peralta
Apparaturen (f) für Peraltamaschine
maquinaria (f) peralta (f)
équipement (m) en matériel peralta
macchinari (m) – peralta

10 machinery – pickers, picking bands, straps, sticks, etc.
Maschinerie (f) für Pickers, Schlagriemen, Gurte, Stöcke usw.
maquinaria (f) para sacar lo sucio del algodón y sus cintas, correas, palos, etc.
matériel (m) – fouets, battants, courroies de chaîne, etc.
macchinari (m) – battitoi, nastri di battitura, nastri, bacchette, ecc.

11 machinery – piece and sewing
Maschinerie (f) für Endstücknäharbeit
maquinaria (f) para coser los bordes de las piezas
matériel (m) – à coudre les bouts
macchinari (m) – cucitura del principio della pezza

12 machinery – pinking
Maschinerie (f) Zum Auszacken
maquinaria (f) cortadora de bordes adentellados
matériel (m) – à découper les bords
macchinari (m) – decorazioni a piccoli fori

1 machinery – pins
Maschinen (f pl) – Bolzen (m pl)
maquinaria (f) de alfileres (f pl)
équipement (m) en matériel – picots (m pl)
macchinari (m) – spilli/spine (f) aghi di filatura

2 machinery – pirns
Maschinen (f pl) – Rollen (f pl)
maquinaria (f) de canillas
équipement (m) en matériel – canettes (f pl)
macchinari (m) – cannette – rocchetti (m pl) Spole (f pl)

3 machinery – pirn winding
Maschinen (f pl) für Rollenwickler
maquinaria (f) de desvanar canillas
équipement (m) en matériel – canetières (f pl)
macchinari (m) – confezione delle cannette/rocchetti

4 machinery – plaiting machines
Apparaturen (f pl) für Flechtmaschinen
maquinaria (f) de plegadores
équipement (m) en matériel à tortiller/à dosser
macchinari (m) – faldatrici

5 machinery – pneumatics
Maschinen (f pl) für Drucklufteinrichtungen
maquinaria (f) neumática (f)
équipement (m) en matériel pneumatique
macchinari (m) – pneumatico

6 machinery – polishing
Maschinerie (f) Zum Polieren
maquinaria (f) de pulir
matériel (m) – à polir
macchinari (m) – lucidatura (f)

7 machinery – polypropylene extrusion
Maschinen (f pl) für Polypropylen-strangpresse
maquinaria (f) de estirar polipropileno a presión
équipement (m) en matériel – extrusion (f) de polypropylène
macchinari (m) – estrusione di polipropilene

8 machinery – presses – balling
Maschinen (f pl) für Ballenpresse
maquinaria (f) de prensas para embalar
équipement (m) en matériel – presses (f pl) à emballer
macchinari (m) – presse – per imballaggio

9 machinery – presses – bump
Maschinen (f pl) für Walkpresse
maquinaria (f) de prensar para superficies irregulares
équipement (m) en matériel – presses (f pl) pour doublier écru
macchinari (m) – presse – per superfici irregolari

10 machinery – rag cleaning and shaking
Maschinerie (f) für Lumpenreinigung und Entstauben
maquinaria (f) para limpiar y sacudir trapos
matériel (m) – à nettoyer et à battre les chiffons
macchinari (m) – pulitura e agitazione degli stracci

11 machinery – rag tearing
Maschinen (f pl) für Lumpen-reißwolf
maquinaria (f) de arrancar trapos
équipement (m) en matériel – effilocheuses (f pl) pour chiffons
macchinari (m) – sfilacciatura degli stracci

12 machinery – raising
Rauhmaschinen (f) – Apparaturen (f pl)
maquinaria (f) de percher
équipement (m) en matériel – égratigneuses (f pl)
macchinari (m) – garzatura

13 machinery – reconditioned
renovierte Maschinen (f pl)
maquinaria (f) revisada
équipement (m) en matériel – reconditionné
macchinari (m) – ricondizionato

14 machinery – reeds
Maschinen (f pl) für Riedblätter
maquinaria (f) de peines
équipement (m) en matériel – guide-fils (m pl)
macchinari (m) – pettini

1 machinery – reels
Hospel (f) – Gerätschäften (f pl)
maquinaria (f) de devanaderas
équipement (m) en matériel – dévidoirs (m pl)
macchinari (m) – aspe (f pl) aspi (m pl)

2 machinery – removers and repairs
Maschinen (f pl) für Reparatur- und Abtransport
cambios (f pl) y reparaciones (f pl) de maquinaria
équipement (m) en matériel – déménageurs (m pl) et réparateurs (m pl)
macchinari (m) – rimuovere e riparare

3 machinery – ring frame
Apparaturen (f) für Ringspinn-maschine
maquinaria (f) hiladora de anillo
équipement (m) en matériel métiers (m pl) continus
macchinari (m) – filatoio ad anelli

4 machinery – ring spinning
Maschinen (f pl) für Ringspinnen
maquinaria (f) hiladora de anillo
équipement (m) en matériel de filage à anneau
macchinari (m) – filatoio ad anelli

5 machinery – ring travellers
Maschinen (f pl) für Ringläufer
maquinaria (f) de anillos móviles
équipement (m) en matériel – curseurs (m pl)
macchinari (m) – elementi mobili del filatoio ad anelli (filatura ad anelli)

6 machinery – ring twisting
Maschinen (f pl) für Ringzwirn (m)
maquinaria (f) de retorcer por anillo
équipement (m) en matériel – continus (m pl) à retordre à anneau
macchinari (m) – ritorcitoio ad anelli

7 machinery – roller cots
Maschinerie (f) für Walzenbezug
manguitos (m) para rodillos
matériel (m) – porte cylindre
macchinari (m) – rivestimenti dei cilindri

8 machinery – roller covers
Maschinerie (f) für Walzenab-deckung
maquinaria (f) para cubiertas de rodillos
matériel (m) – chapeaux de cylindre
macchinari (m) – coperture per rulli

9 machinery – rollers
Maschinen (f pl) für Walzen /Rollen (f pl)
maquinaria (f) de rodillos
équipement (m) en matériel – rouleaux (mpl)
macchinari (m) – cilindri

10 machinery – roller – batching
Maschinerie (f) für Rollbatschen
maquinaria (f) de rodillos de agrupación
équipement (m) en matériel – rouleaux (m pl) d'enroulage
macchinari (m) – cilindri – oliaggio

11 machinery – rollers heated
Maschinerie (f) für beheizte Rollen
maquinaria (f) de rodillos calentados
équipement (m) en matériel – rouleaux (m pl) préchauffés
macchinari (m) – cilindri caldi

12 machinery – rope detwisters
Maschinerie (f) für Seilaufdreher
maquinaria (f) para desenrollar sogas
matériel (m) – à détendre les boyaux
macchinari (m) – distenditori di corde

13 machinery – ropes
Maschinerie (f) für Seilerei
maquinaria (f) para cuerdas
équipement (m) en matériel – boyaux (m pl)
macchinari (m) – corde

14 machinery – rotor open end spinning
Maschinerie (f) für Rotorendlos-spinnanlage/Open-End Spinnen (n)
maquinaria (f) de hilar por rotores de extremo abierto
équipement (m) en matériel – filature (f) à fibres libérées à rotor
macchinari (m) – rotore – filatura open-end

1 machinery – roving
Maschinen (f pl) für Vorgarnein-richtung
maquinaria (f) de primera torsión
équipement (m) en matériel – pour mèches
macchinari (m) – lucignolo/lucignolo di banco a fusi

2 machinery – rubber aprons
Maschinen (f pl) – für Mitschelgummi
maquinaria (f) para planchas de protección de goma
équipement (m) en matériel – tabliers (m pl) en caoutchouc
macchinari (m) – nastri trasportatori di gomma

3 machinery – rubber cots
Maschinen (f pl) für Gummi walzenbezüge
maquinaria (f) para manguitos de goma
équipement (m) en matériel – garnitures (f pl) de rouleaux en caoutchouc
macchinari (m) – manicotti di rivestimento in gomma

4 machinery – scouring
Apparaturen (f pl) für Putzmaschine
maquinaria (f) de desengrasar
équipement (m) en matériel – pour le lavage
macchinari (m) – purga

5 machinery – scouring – loose wool
Maschinerie (f) für Waschen, lose Wolle
maquinaria (f) para desengrasar lana suelta
matériel (m) – lavage de laine en bourre
macchinari (m) – purga di lana sciolta

6 machinery – second hand
Apparaturen (f pl) für Gebrauchtmaschinen /gebrauchte Anlage (f)
maquinaria (f) de ocasión
équipement (m) en matériel d'occasion
macchinari (m) – usato/di seconda mano

7 machinery – selvedge uncurlers
Maschinerie (f) für Kantenent-kräusler, Leistenausroller (m)
maquinaria (f) para desrizar hirmas
matériel (m) – dérouleurs de lisière
macchinari (m) – anti increspatura cimosa

8 machinery – servicing
Maschinen (f pl) für Überholen
servicio (m) de maquinaria
équipement (m) en matériel – d'entretien
macchinari (m) – assistenza

9 machinery – shearing
Maschinen (f pl) für Scheranlage
maquinaria (f) para tundir
équipement (m) en matériel – tondeuses (f pl)
macchinari (m) – cimatura

10 machinery – shrinking
Maschinen (f pl) für Schrumpfanlage
maquinaria (f) para encoger
équipement (m) en matériel – de rétrécissement
macchinari (m) – restringimento

11 machinery – shuttles and accessories
Maschinen (f pl) für Schüften (Schiffchen) und Zusatzteile
maquinaria (f) de lanzaderas y accesorios
équipement (m) en matériel – navettes (f pl) et accessoires (m pl)
macchinari (m) – spole ed accessori

12 machinery – singeing
Apparaturen (f pl) für Sengen, Gasieren (n)
maquinaria (f) de chamuscar
équipement (m) en matériel à flamber
macchinari (m) – gasatura

13 machinery – sizing
Apparaturen (f pl) für Schlichtmaschine
maquinaria (f) de encolar
équipement (m) en matériel – à encoller
macchinari (m) – imbozzimatura

1 machinery – spares
Maschinen (f pl) – Ersatzteile (n pl)
repuestos (m pl) de maquinaria
équipement (m) en matériel – pièces (f pl) de rechange
macchinari (m) – pezzi di ricambio

2 machinery – special purpose
Apparaturen (f pl) für Spezialzweckmaschinen
maquinaria (f) para trabajos especiales
équipement (m) en matériel – spécialisé
macchinari (m) – per uso speciale

3 machinery – spinning
Apparaturen (f pl) für Spinnmaschine
maquinaria (f) de hilar
équipement (m) en matériel – métiers (m pl) à filer
macchinari (m) – filatura

4 machinery – spinning and twisting rings
Maschinerie (f) für Spinn- und Zwirnringe
maquinaria (f) de hilar y torcer por anillos
matériel (m) – anneaux de filage et de retordage
macchinari (m) – anelli di filatura e torsione

5 machinery – spinning – hollow spindle
Maschinerie (f) für Hohlspindel-Spinnen
maquinaria (f) de hilar por huso hueco
matériel (m) – filature à broche creuse
macchinari (m) – fuso cavo

6 machinery – spinning – open end
Maschinerie (f) für Open-End-Spinnen
maquinaria (f) de hilar – extremo abierto
matériel (m) – filature à fibres libérées
macchinari (m) – filatura open-end

7 machinery – spinning rings
Maschinerie (f) für Spinn-Ringe
anillos (m pl) de maquinaria de hilar
matériel (m) – filature – anneaux
macchinari (m) – anelli

8 machinery – spinning travellers
Maschinerie (f) für Spinn-Ringläufer
correderas (f pl) de maquinaria de hilar
matériel (m) – filature – curseurs
macchinari (m) – elementi mobili

9 machinery – spinning – wet ring
Maschinerie (f) für Naßring-Spinnen
maquinaria (f) de hilar por anillo humedo
matériel (m) – filature au mouillé
macchinari (m) – anello per ritorcitoio a umido

10 machinery – spinning worsted
Maschinerie (f) für Kammgarn-Spinnen
maquinaria (f) de hilar estambre
matériel (m) – filature de laine peignée
macchinari (m) – pettinato

11 machinery – spread rolls
Maschinerie (f) für Breithalterwalzen
rodillos (m pl) para repartir urdimbre
matériel (m) – rouleaux d'imprimeur
macchinari (m) – rulli di allargatore

12 machinery – squeeze press
Maschinerie (f) für Quetschpressen
maquinaria (f) de prensar para exprimir
matériel (m) – foulards-exprimeurs
macchinari (m) – pressa spremitrice

13 machinery – stenter
Maschinen (f pl) für Spannrahmen
maquinaria (f) stenter
équipement (m) en matériel – rames (f pl)
macchinari (m) – ramense (f)

14 machinery – stenter guides
Maschinerie (f) für Spannführungen [Malimo-Verfahren]
maquinaria (f) – guías stenter
matériel (m) – guides de rameuses
macchinari (m) – guide rameuse

15 machinery – stitch bonding
Maschinerie (f) für Nähwirken [Malimo-Verfahren]
maquinaria (f) de coser con puntos
matériel (m) – Malimo
macchinari (m) – fissaggio mediante cucitura

1 machinery – stop motion
Maschinerie (f) für Absteller
maquinaria (f) de parar el movimiento
matériel (m) – dispositifs de casse-ruban
macchinari (m) – arresto automatico

2 machinery – stop motions – knitting
Maschinen (f pl) für Ausrückvorrichtung
aparatos (m pl) de interrumpir el funcionamiento de maquinaria de hacer punto
équipement (m) en matériel – dispositifs (m pl) de casse-ruban pour tricots
macchinari (m) – arresto automatico – maglieria

3 machinery – stroboscopes and tachometers
Maschinerie (f) als Stroboskop und Tachometer
maquinaria (f) – estroboscopios y tacómetros
matériel (m) – stroboscopes et compteurs de tours
macchinari (m) – stroboscopi e tachimetri

4 machinery – sueding
Maschinerie (f) für Velourieren
maquinaria (f) de dar aspecto de ante
matériel (m) – émerisseuses
macchinari (m) – per pelle scamosciata

5 machinery – swab making
Maschinerie (f) für Scheuerlappenherstellung (f)
maquinaria (f) para fabricar escobones
matériel (m) – fabrication de torchons
macchinari (m) – per spazzole

6 machinery – take up roller covering
Maschinerie (f) für Aufspulwalzenbezug
maquinaria (f) para recoger cubiertas de rodillos
matériel (m) – courverture d'enrouleur
macchinari (m) – copertura per rullo di avvolgimento

7 machinery – teazles
Maschinen (f pl) für Rauhkarden
maquinaria (f) para cardenchas
équipement (m) en matériel de lainage
macchinari (m) – cardi per garzatura

8 machinery – tension meters
Maschinerie (f) für Spannungsmesser
maquinaria (f) para medir tensiones
matériel (m) – mesure de tension
macchinari (m) – contatori di tensione

9 machinery – tenterhook willeys
Maschinerie (f) für Spanngreiferreißwölfe
maquinaria (f) para limpiar ganchos de tendedores
matériel (m) – loups à pince
macchinari (m) – diavolotti a gancio per asciugatura

10 machinery – texturing
Maschinerie (f) für Texturieren
maquinaria (f) para dar textura
matériel (m) – texturation
macchinari (m) – tessitura

11 machinery – thermal bonding
Maschinerie (f) für Hitzeverkleben
maquinaria (f) de unir por calor
matériel (m) – liage thermique
macchinari (m) – legame termico

12 machinery – thermasol dyeing
Maschinerie (f) für Thermasol färben
maquinaria (f) para teñir estilo termosol
matériel (m) – teinture thermosolage
macchinari (m) – tintura thermasol

13 machinery – thickness gauges
Maschinerie (f) für Stärke messer
calibradores (m pl) de espesor
matériel (m) – jauges d'épaisseur
macchinari (m) – misuratori dello spessore/calibri dello spessore

14 machinery – tigering
Maschinerie (f) für Rauhen
maquinaria (f) de separación
matériel (m) – lainage
macchinari (m) – per disegni di riproduzione del pelo delle tigri

1 machinery – top making
Maschinerie (f) für Kammzug-fertigung
maquinaria (f) de fabricar husos
matériel (m) – fabrication de ruban peignés
macchinari (m) – per nastri

2 machinery – tow- and sliver-cutting
Maschinerie (f) für Tow- und Spinnbandschneiden
maquinaria (f) de cortar estopa y mechas
matériel (m) – à couper les câbles et les rubans de fibres
macchinari (m) – per tagliare stoppa e nastri di carda

3 machinery – tow convertors
Maschinen (f pl) für Spinnkabelver-arbeiter
maquinaria (f) para convertir estopa
équipement (m) convertisseurs (m pl) pour câble
macchinari (m) – converter di stoppa

4 machinery – transfer printing
Maschinen (f pl) für Transfer-druckanlage
maquinaria (f) para transferir estampas
équipement (m) en matériel d'impression hectographique
macchinari (m) – stampa a decalcomania

5 machinery – transmission
Maschinen (f pl) der Antriebsüber-tragungen
maquinaria (f) para transmisiones
équipement (m) en matériel – transmissions (f pl)
macchinari (m) – trasmissione

6 machinery – trucks and trolley manufacturers
Maschinerie (f) für Transport Wagenherstellung
fabricantes (m pl) de carros y carretillas
matériel (m) – fabricants de chariots et fardiers
macchinari (m) – produttori di autocarri e carrelli

7 machinery – tubes
Maschinerie (f) als Rohre /Röhren (f pl)
maquinaria (f) para tubos
équipement (m) en matériel – tubes (m pl)
macchinari (m) – per tubi

8 machinery – tubular finishing
Maschinerie (f) für Schlauchappretur
maquinaria (f) de remate tubular
matériel (m) – finissage tubulaire
macchinari (m) – finissaggio tubolare

9 machinery – tufting
Maschinen (f pl) für Tufting
maquinaria (f) para tufting
équipement (m) en matériel – au tufting
macchinari (m) – tufting

10 machinery – twisting
Apparaturen (f pl) für Zwirnmaschine
maquinaria (f) de retorcer
équipement (m) en matériel à retordre
macchinari (m) – ritorcitura

11 machinery – uptwisters
Maschinen (f pl) für Etagenzwirnen
maquinaria (f) de retorcer de paso ascendente
équipement (m) en matériel – retordeuses (f pl) à plusieurs rangées de bobines
macchinari (m) – torcitoi a piani

12 machinery – velvet finishing
Maschinerie (f) für Samtappretur
maquinaria (f) para acabar terciopelo
matériel (m) – apprêt du velours
macchinari (m) – finissaggio di velluto

13 machinery – warp drawing-in
Maschinen (f pl) für Ketteneinzieh-vorrichtung
maquinaria (f) de tirar urdimbre
équipement (m) en matériel à rentrer les chaînes
macchinari (m) – rimettaggio della trama

1 machinery – warp sizing
Apparaturen (f pl) für Kettenschlichtmaschine
maquinaria (f) de aprestar urdimbre
équipement (m) en matériel d'encollage de la chaîne
macchinari (m) – imbozzimatura della trama

2 machinery – warp stop motion
Maschinen (f pl) für Kettenabstell-vorrichtung
maquinaria (f) de interrumpir el recorrido de urdimbre
équipement (m) en matériel – casse-chaînes (m pl)
macchinari (m) – guardaordito (m)

3 machinery – warp tying
Apparaturen (f pl) für Webkettenanknüpfmashine better: Webketten-A.
maquinaria (f) de ligar urdimbre
équipement (m) en matériel – noueuses (f pl) de chaîne
macchinari (m) – annodatura della trama

4 machinery – warping
Apparaturen (f pl) für Schärmaschine
maquinaria (f) de urdido
équipement (m) en matériel – ourdissoirs (m pl)
macchinari (m) – orditura

5 machinery – warping accessories
Maschinen (f pl) als Schärzusatateile
maquinaria (f) de accesorios de urdido
équipement (m) en matériel – accessoires (m pl) d'ourdissage
macchinari (m) – accessori di orditura

6 machinery – warping creels
Maschinen (f pl) für Schärgatter
maquinaria (f) de cestas de urdido
équipement (m) en matériel – centres (m pl) d'ourdissage
macchinari (m) – cantre di orditura

7 machinery – washing
Apparaturen (f pl) für Waschmaschine
maquinaria (f) de lavar
équipement (m) en matériel à laver
macchinari (m) – lavaggio

8 machinery – waste or fibre cleaning or shaking
Maschinen (f pl) für Abfall- oder Fasersäuberungs- oder Schüttelvorrichtung
maquinaria (f) para limpiar o sacudir residuos o fibras
équipement (m) en matériel – à nettoyer ou à secouer les déchets ou les fibres
macchinari (m) – cascame, pulitura fibre od agitazione

9 machinery – waste- or fibre-cutting
Maschinerie (f) für Ausschuß- oder Faserschneiden
maquinaria (f) para cortar residuos o fibras
matériel (m) – à découper les déchets ou les fibres
macchinari (m) – cascame/scarto o traglio di fibre

10 machinery – weaving frames
Maschinen (f pl) für Webrahmen
maquinaria (f) de telas
équipement (m) en matériel – métiers (m pl) à tisser
macchinari (m) – telai

11 machinery – web
Maschinen (f pl) für Netz /Fasergarn, Faserflor
maquinaria (f) de tela tejida y rematada
équipement (m) en matériel – plieurs (m pl) de voile
macchinari (m) – velo/velo di carda

12 machinery – web forming
Maschinerie (f) für Florbildung
maquinaria (f) para formar capas de fibras no hiladas
matériel (m) – à former un voile
macchinari (m) – formazione del velo

13 machinery – web guiding
Maschinerie (f) für Florführung
maquinaria (f) para guiar fibras no hiladas
matériel (m) – à guider un voile
macchinari (m) – guida del velo

1 machinery – web spreading
Maschinerie (f) für Florausbreitung
maquinaria (f) para tender fibras no hiladas
matériel (m) – à plier un voile
macchinari (m) – propagazione di trama

2 machinery – web tension
Maschinerie (f) für Florspannung
maquinaria (f) para tensionar fibras no hiladas
matériel (m) – tension du voile
macchinari (m) – tensione di trama

3 machinery – web winding drums
Maschinerie (f) für Florwickeltrommeln
tambores (m pl) para devanar fibras no hiladas
matériel (m) – cylindres à enrouler un voile
macchinari (m) – tamburi avvolgitori di trama

4 machinery – web woollencard accessories
Maschinerie (f) für Flor-Wollkarden-Zubehör
accesorios (m pl) para maquinaria de capas de fibras de lana cardada
matériel (m) – accessoires de voile de laine cardée
macchinari (m) – accessori per trama con carda di lana

5 machinery – weft forks
Maschinen (f pl) für Kuliergabeln
maquinaria (f) de horquillas de trama
équipement (m) en matériel – fourchettes (f pl) de casse-trame
macchinari (m) – forcelle di trama

6 machinery – weft straighteners
Maschinen (f pl) für Kulierstrecker
maquinaria (f) de enderezar trama
équipement (m) en matériel – redresseurs (m pl) de trame
macchinari (m) – raddrizzatrici della trama

7 machinery – web yarn control equipment
Maschinerie (f) für Florgarnkontrollgeräte
maquinaria (f) para revisar hilo de fibras no hiladas
matériel (m) – équipement de commande du fil poil
macchinari (m) – apparecchiature per il controllo delle trame

8 machinery – web yarn measuring
Maschinerie (f) für Florgarnmessung
maquinaria (f) para medir hilo de fibras no hiladas
matériel (m) – mesure du fil poil
macchinari (m) – misurazione di trama

9 machinery – web yarn monitoring
Maschinerie (f) für Florgarnüberwachung
maquinaria (f) para controlar fibras no hiladas
matériel (m) – contrôle du fil poil
macchinari (m) – misurazione di trama

10 machinery – web yarn printing
Maschinerie (f) für Florgarndrucken
maquinaria (f) para estampar hilo de fibras no hiladas
matériel (m) – impression du fil poil
macchinari (m) – stampaggio di trama

11 machinery – web yarn relaxing
Maschinerie (f) für Florgarnentspannung
maquinaria (f) para aflojar hilo de fibras no hiladas
matériel (m) – relaxation du fil poil
macchinari (m) – allungamento di trama

12 machinery – web yarn washing and drying
Maschinerie (f) für Florgarnwaschen und -trocknen
maquinaria (f) para lavar y secar hilo de fibras no hiladas
matériel (m) – lavage et séchage du fil poil
macchinari (m) – lavaggio ed asciugatura delle trame

1 machinery – web yarn mothproofing
Maschinerie (f) für Florgarnmotten-schutz
maquinaria (f) para protección contra polillas de hilo de fibras no hiladas
matériel (m) – apprêt antimite du fil poil
macchinari (m) – trattamento anti-tarme di trama

2 machinery – width measurement
Maschinen (f pl) für Breitenmeßein-richtung
maquinaria (f) de medir el ancho
équipement (m) mesure de la largeur
macchinari (m) – misura della larghezza

3 machinery – winding
Maschinen (f pl) für Aufwickel-vorrichtung
maquinaria (f) de hilar
équipement (m) en matériel à renvider
macchinari (m) – incannatura

4 machinery – wool cleaning or shaking
Maschinen (f pl) für Wollsäuberungs- oder Schüttelanlage
maquinaria (f) para limpiar o sacudir lana
équipement (m) en matériel à nettoyer ou à secouer la laine
macchinari (m) – pulitura o agitazione di lana

5 machinery – wrap spinning
Apparaturen (f pl) für Umwindespinn-maschine (f)
maquinaria (f) de hilar wrap
équipement (m) en matériel à filer au guipage
macchinari (m) – filatura ad avvolgimento

6 machinery – yarn clearers
Maschinen (f pl) für Garnreiniger-vorrichtung
maquinaria (f) de desenredar hilo
équipement (m) en matériel – purgeurs (m pl) de fil
macchinari (m) – stribbie (f pl)

7 machinery – yarn conditioners
Maschinen (f pl) für Garnkonditio-nierer
maquinaria (f) de acondicionar hilo
équipement (m) en matériel – de conditionnement du fil
macchinari (m) – umidificazione dei filati

8 machinery – yarn guides
Maschinen (f pl) für Farnführungs-schienen
maquinaria (f) de guías de hilo
équipement (m) en matériel – guichet-fils (m pl)
macchinari (m) – guide del filato

9 machinery – yarn heat setting
Maschinen (f pl) für Farnheißfixier-ungsanlage
maquinaria (f) de fijar hilo por termofijación
équipement (m) en matériel de thermofixage de fil
macchinari (m) – filato – regolazione del calore

10 machinery – yarn joining
Maschinen (f pl) für Garnver-bindungen
maquinaria (f) de ligar hilos
équipement (m) en matériel de rattachement de fil
macchinari (m) – unione del filato

11 machinists
Maschinist (m)
maquinistas (m pl)
machinistes (m pl)
macchinisti (m pl)

12 maintenance
Wartung (f)
manutención/mantenimiento (m)
entretien (m)
manutenzione (f)

13 mangle
Mangel (f)
planchadora (f)
exprimeuse
mangano (m)

14 manmade
künstlich (adj)
sintético
synthétique
sintetico

1 manmade fibres
Kunstfasern (f pl)
fibras (f pl) sintéticas
fibres (f pl) synthétiques
fibre (f pl) sintetiche/chimiche

2 manmade yarn spinners
Garnspinner (m pl) ans Kunstafasern
hiladores (m pl) de hilo sintético
filateurs (m pl) de fil synthétique
filatori di filato sintetico (f pl)

3 manufacturer
Hersteller (m)
fabricante (m)
fabricant (m)
produttore (m)

4 market
Markt (m)
mercado (m)
marché
mercato (m)

5 market research
Marktforschung (f)
estudios de mercado
étude (m) de marché
ricerca (f) di mercato

6 marketing
Marketing (n)
ventas (f pl)
marketing, commercialisation (m)
marketing (m)/commercio/mercatura

7 material
Material (n)
material (m)
tissu (m)/étoffe (f)
stoffa (f)

8 materials handling systems
Materialbehandlungssysteme (n pl)
sistemas (m pl) de manipulación de materiales
systèmes (m pl) de manutention
sistemi (m pl) di movimentazione dei materiali

9 measure
Maß (n)
medida (f)
mesurer (v)
misura (f)

10 mechanic
Mechaniker (m); mechanisch (adj)
mecánico (m)
mécanicien (m)
meccanico (agg)

11 mechanical
mechanisch (adj)
mecánico
mécanique
meccanico

12 mechanism
Mechanismus (m)
mecanismo (m)
mécanisme (m)
meccanismo (m)

13 melting
Schmelzen (n)
fusión (f)
fusion (f)
fusione (f)

14 mending [cloth]
Ausbessern (n)
reparación (f) remiendo (m)
raccommodage (m)
rammendo (m) [tessuto]

15 mercerizing
Merzerisieren (n)
mercerizar (m)
mercerisage (m)
mercerizzazione (f)

16 merchant
Händler (m)
comerciante (m)
négociant (m)
commerciante (m)

17 merchant converter
Händler Verwerter (m)
convertidor (m) mercantil
marchand-convertisseur (m)
convertitore (m) commerciale

18 metre
Meter (m)
metro (m)
mètre (m)
metro (m)

1 metric
metrisch (adj)
métrico
métrique
metrico

2 metric count
metrische Nummer (f), Nin (abbr)
cuenta métrica (f)
numéro (m) métrique
conteggio (m) metrico, titolo metrico

3 mildew
Schimmel (m)
moho (m)
moisissure (f)
muffa (f)

4 mill [building]
Fabrik (f)
fábrica (f)
usine (f) textile
[edificio] fabbrica (f)

5 mill furnishers
Fabriksausstatter (m pl)
suministradores (m pl) para fábricas
fournisseurs (m pl) d'usine textile
fornitori (m pl) di fabbrica

6 milling [cloth]
Walken (n), Foulieren (n)
abatanado
foulage (m)
follatura

7 mineral dye
Mineralfarbstoff (m)
tinte (f) mineral
teinture (f) minérale
tintura (f) minerale

8 mineral oil
Mineralöl (n)
aceite (m) mineral
huile (f) minérale
olio (m) minerale

9 mixture
Mischung (f)
mezcla (f)
mélange (m)
miscela (f)

10 modernise, modernize (v)
modernisieren
modernizar
moderniser
modernizzare

11 modification
Modifikation (f)
modificación (f)
modification (f)
modifica (f)

12 mohair scoured
Mohair (m) – gewaschen
pelo (m) de angora desengrasado
mohair (m) lavé à fond
mohair (m) lavato

13 moisture
Feuchtigkeit (f)
humedad (f)
humidité (f)
umidità (f)

14 moisture content
Feuchtigkeitsgehalt (m)
porcentaje (m) de humedad
teneur (f) en eau, pourcentage d'eau
indice/livello (m) di umidità/tenore di umidità

15 moisture meters
Feuchtigkeitsmesser (m)
contadores (m pl) de humedad
appareils (m pl)O de mesure d'humidité
contatori di umidità

16 molecular
molekular (adj)
molecular
moléculaire
molecolare

17 moth
Motte (f)
polilla (f)
mite (f)
tarma (f)

18 mothproof
mottenecht (adj)
protegido contra polillas
antimite
antitarme

1 mothproofing agents
Mottenschutzmittel (n pl)
agentes (m pl) de protección contra polillas
antimites (m pl)
agenti (m pl) antitarme

2 multicolour
vielfarben (adj)
multicolor
multicolore
multicolore

3 mungo
Lumpenreißwolle (f), Mungo (m)
fibras (f pl) reexplotadas de lana
laine (f) renaissance
lana (f) a fibra corta mungo

4 mungo and shoddy manufacturers
Mungo- und Shoddyhersteller(m pl), Regeneratwollhersteller (m pl)
fabricantes (m pl) de fibras reexplotadas de lana
fabricants (m pl) de laine renaissance et de shoddy
produttori (m pl) di lana a fibra corta mongo e di lana rigenerata

N

5 narrow fabric
Bandware (f), Schmalgewebe (n)
tela (f) estrecha
tissu (m) étroit
tessuto (m) basso

6 narrow fabrics
Bandwaren (f pl), Schmalgewebe (n pl)
telas (f pl) estrechas
rubans (m pl)
tessuti (m pl) bassi

7 napping
Aufrauhen (n)
perchar (m)
grattage (m)
garzatura (f)

8 natural fibre
Naturfaser (f)
fibra (f) natural
fibre (f) naturelle
fibra (f) naturale

9 needle
Nadel (f)
aguja (f)
aiguille (f)
ago (m)

10 needle loom
Nadelwebstuhl (m)
tejar de agujas (m pl)
aiguilleteuse (f)
telaio (m) ad aghi

11 needles
Nadeln (f pl)
agujas (f pl)
aiguilles (f pl)
aghi (m pl)

12 nep
Knoten (m), Knötchen (n)
nudo (f) de fibritas enredadas de algodón
neps (m pl)
nodulo (m) nep

13 nett
netto
neto
net
netto

14 nett weight
Nettogewicht (n)
peso (m) neto
poids (m) net
peso (m) netto

15 neutral
neutral (adj)
neutral
neutre
neutro

16 neutralise, neutralize (v)
neutralisieren
neutralizar
neutraliser
neutralizzare

17 nitrate
Nitrat (n)
nitrato (m)
nitrate (m)
nitrato (m)

1 nitric acid
Salpetersäure (f)
acido (m) nítrico
acide (m) nitrique
acido (m) nitrico

2 noble comb
Noble-Wollkamm (m)
peine (m) noble
peigneuse (f) Noble
pettine (m) nobile

3 noils
Kämmlinge (m pl)
fibritas (f pl) peinadas
blousse (f)
cascame (m) di pettinatura

4 noils – alpaca
Alpakakämmlinge (m pl)
fibritas (f pl) peinadas de alpaca
blousse (f) d'alpaga
cascame (m) di pettinatura-
alpaca

5 noils – angora
Angorakämmlinge (m pl)
fibritas (f pl) peinadas de angora
blousse (f) d'angora
cascame (m) – di pettinatura-
angora

6 noils – camel hair
Kamelhaarkämmlinge (m pl)
fibritas (f pl) peinadas de pelo de
camello
blousse (f) de poil de chameau
cascame (m) di pettinatura –
pelo di cammello

7 noils – carbonised, carbonized
karbonisierte Kämmlinge (m pl)
fibritas (f pl) peinadas y cachemir
blousse (f) carbonisée
cascame (m) – di pettinatura –
carbonizzato

8 noils – cashmere
Kaschmirkämmlinge (m pl)
fibritas (f pl) peinadas de casimir
blousse (f) de cachemire
cascami (m pl) – cashmere

9 noils – coloured
gefärbte Kämmlinge (m pl)
fibritas (f pl) peinadas de colores
blousse (f) colorée
cascami (m pl) – colorati

10 noils – crossbred
Kreuzzuchtkämmlinge (m pl)
fibritas (f pl) peinadas y
mezcladas
blousse (f) croisée
cascami (m pl) – lana incrociata

11 noils – goat hair
Ziegenhaarkämmlinge (m pl)
fibritas (f pl) peinadas de pelo
de cabra
blousse (f) de poil de chèvre
cascami (m pl) – pelo di capra

12 noils – manmade
Kunstfaserkämmlinge (m pl)
fibritas (f pl) sintéticas
blousse (f) synthétique
cascami (m pl) – artificiali

13 noils – merino
Merinoschafkämmlinge (m pl),
Merino-Kämmlinge (m pl)
fibritas (f pl) peinadas merino
blousse (f) mérinos
cascami (m pl) – merino

14 noils – mohair
Mohairkämmlinge (m pl)
fibritas (f pl) peinadas de pelo de
angora
blousse (f) de mohair
cascami (m pl) – mohair

15 noils – polyester
Polyesterkämmlinge (m pl)
fibritas (f pl) peinadas de
poliéster
blousse (f) de polyester
cascami (m pl) – poliestere

16 noils – silk
Seidenkämmlinge (m pl)
fibritas (f pl) peinadas de seda
blousse (f) de soie, bourrette (f)
cascami (m pl) – seta

17 noils – wool
Wollkämmlinge (m pl)
fibritas (f pl) peinadas de lana
blousse (f) de laine
cascami (m pl) – lana

18 noils – yak
Yakkämmlinge (m pl)
fibritas (f pl) peinadas de yak
blousse (f) de yak
cascami (m pl) – yak

1 nops – wool/synthetic
Noppen (f pl) – Wolle Synthetik
nudos (m pl) de lana/sintético
noppes (f pl) – laine/synthétique
nodi (m pl) di lana/sintetici

2 novelty
Neuigkeit (f)
novedad (f)
fil (m) fantaisie/nouveauté (f)
novità (f)

3 nylon
Nylon (n)
nilón (m)
nylon (m)
nylon (m)/nailon

4 nylon fibre
Nylonfaser (f)
fibra (f) de nilón
fibre (f) nylon
fibra (f) di nailon

5 nylon staple
Nylonkurzfaser (f), Nylon-stapelfaser (f)
filamento (m) de nilón
soie (f) de nylon
fiocco di nailon/di Nylon

O

6 Oil
Öl (n)
aceite (m)
huile (f)
olio (m)

7 Oiling
Ölung (f)
lubrificar (m)
ensimage (m)
oliatura (f)

8 Oils – anti-static
Antistatiköle (n pl)
aceites (m pl) antiestáticos
huiles (m pl) antistatiques
oli (m pl) – antistatici

9 Oils – coning
Spulenöle (n pl)
aceites (m pl) de devanar
lubrifiants (m pl) de bobinage
oli (m pl) – per rocchetti

10 Oils – fuel
Öle (n pl) – Brennstoffe (m pl)/ Treibstoffe (mpl)
combustibles (m pl)
combustibles (m pl)
oli (m pl) – combustibili

11 Oils – lubricants
Öle (n pl) – Gleitmittel (n pl)
lubrificantes (m pl)
lubrifiants (m pl)
oli (m pl) – lubrificanti

12 Oils – spray
Öle (n pl) – Sprays (n pl)
aceites (m pl) de envase a presión
huiles (f pl) de pulvérisation
oli (m pl) – spray

13 Olein
Olein (n)
oleina (f)
oléine (f)
oleina (f)

14 oleines
Oleine (n pl)
oleínas (f pl)
oléines (f pl)
oleine (f pl)

15 opaque
opak, undurchsichtig (adj)
opaco
opaque
opaco

16 open end spinning
Endlosspinnen (n), Open-End–Spinnen (n), OE-Spinnen (n)
hilatura (m) de extremo abierto
filature (f) à fibres libérées
filatura (f) ad estremità aperta/open end

17 opener
Öffner (m), Wolf (m)
desmotadora (f)
ouvreuse (f)
apritoio (m)

1 openers – synthetic
Wölfe (m pl) – Synthetikwolf (m)
desmotadoras (f pl) de sintético
effilocheuses (f pl) pour synthétiques
apritoi (m pl) – per sintetici

2 openers – waste
Wölfe (m pl) – Abfallwolf (m)
desmotadoras (f pl) de residuos
effilocheuses (f pl) pour déchets
apritoi (m pl) – per cascame

3 openers – wool
Wölfe (m pl) – Wollwolf (m)
desmotadores (f pl) de lana
effilocheuses (f pl) pour laine
apritoi (m pl) – per lana

4 open shed [weaving]
Stehfach (n)
departamento (m) descubierto
pas (m) ouvert
passo (m) aperto [tessitura]

5 open width
offene Breite (f)
anchura (f) abierta
au large
in largo

6 open width dyeing
Breitfärberei (f)
tinte (m) de anchura abierta
teinture (f) au large
tintura (f) in largo

7 open-width scouring
Breitwäscherei (f)
desengrasar (m) de anchura abierta
lavage (m) au large
lavaggio (m) in largo

8 operator
Arbeiter (m), Bedienungsperson (f)
operario (m)
opérateur (m), opératrice (f)
operatore (m)

9 organic acid
organische Säure (f)
ácido (m) orgánico
acide (m) organique
acido (m) organico

10 overall length
Gesamtlänge (f)
largura (f) total
longueur (f) totale
lunghezza (f) totale

11 overall width
Gesamtbreite (f)
anchura (f) total
largeur (f) totale
larghezza (f) totale

12 overdye (v)
überfärben
tinte (m) de repaso
surteindre
applicare un colore su un altro

P

13 package
Paket (n), Garnkörper (m), Garnspule (f)
bulto (m)
bobine (f)
bobina (f)/pacco (m)

14 package dyeing
Packfärben (n), Spulenfärber (m)
teñido en bobina
teinture (f) sur bobine
tintura (f) in bobina

15 package dyers
Packfärber (m pl)
tintoreros (m pl) de bobinas
teinturiers (m pl) sur bobine
tintori (m pl) in bobina

16 packaging material
Verpackungsmaterial (n)
materiales (m pl) de embalar
matières (f pl) d'emballage
materiale (m) per imballaggio

17 packers – containerisation, containerization
Packer (m pl) – Containerpacken (n)
embaladores (m pl) para container
emballeurs (m pl) – conteneurisation
imballatori (m pl) – containerizzazione

1 packers – press packing
Packer (m pl) – Presspacken (n)
embaladores (m pl) por prensa
emballeurs (m pl) – emballage serré
imballatori (m pl) – imballaggio mediante pressa

2 packers – vacuum packing
Packer (m pl) – Vakuumpacken (n)
embaladores (m pl) por vacuo
emballeurs (m pl) – emballage sous vide
imballatori (m pl) – imballaggio sotto vuoto

3 pad
Polster (n), Unterlage (f)
relleno (m)
rembourrage (m), ouatage (m)
foulard

4 pad (v)
auspolstern, wattieren
acolchar, rellenar
rembourrer, foularder
imbottire

5 pallet-wrap, stretch film
Paletteneinwickelung, Streckfolie (f)
membrana (f) extensible para envolver jergones
enveloppement (m) de palettes, film 'stretch'
fascetta (f) da palette, pellicola allungabile

6 paper and kraft squares
Papier- und Kraftbögen (m pl)
cuadros (m pl) de papel e hilo fuerte
carrés (m pl) en papier et papier kraft
carta (f) millimetrata e carta kraft

7 parallel
parallel (adj)
paralelo
parallèle
parallelo

8 part
Teil (m)
parte (f)
partie (f)
parte (f), pezzo (m)

9 pastel
pastell (adj)
pastel
pastel
pastello

10 patent
Patent (n)
patente (f)
brevet (m)
brevetto (m)

11 pattern [clothing]
Muster (n), Stoffmuster (n)
diseño (m)
dessin (m)
motivo (m) [produz, tessuti]

12 pattern books and shade cards
Musterbuch (n) und Schattierungskarten (f pl)
libros (m pl) de diseños y muestrario (f pl) de tintes
livres d'échantillons et cartes (f pl) de nuances
libri (m pl) di modelli e cartelle colori

13 pattern weaver
Musterweber (m)
tejedor (m) de diseños
tisseur (m) de dessins
tessitore (m) di motivi

14 per cent
Prozent (n)
por ciento
pour cent
per cento

15 peg
Stift (m), Zapfen (m)
clavija (f)
bouton (m)
chiodo (m), unghia (f), tampone (m)

16 peralta
Peralta (n)
peralta
peralta
peralta

17 perching [cloth]
Gewebebeschauen (n), Warenschan (f)
perchar (m)
visite (f)
esame (m) del tessuto

1 perforated
perforiert (adj)
perforado
perforé
perforato

2 permanent
haltbar (adj), daũerhaft (adj), blcibend (adj)
permanente
permanent
permanente

3 permanent setting
bleibende Fixierung (f)
fijación (f) permanente
fixage (m) permanent
regolazione (f) permanente

4 pH metres
pH-Meter (m)
metro (pl) de acidez
dispositif (pl) de mesure de pH
misurator (pl) di acidità

5 pick [weft]
Schußfaden (m)
hilo (m) de trama
duite (f)
trama (f) [orditura]

6 picks per cm
Schüsse pro cm
hilos (m pl) de trama por cm
nombre (m) de duites par centimètre
battute (f pl) per cm

7 piece dyeing
Stückfärben (n)
Teñido (m) de piezas
teinture (f) en pièce
tintura (f) in pezza

8 piece glasses and pick counters
Schußzähler (m)
lupas (f pl) para revisar telas y contar hilos
loupes (f pl) pour entures et compteurs de fils
contafili (m pl) e . . . . . contabattuti (m pl)

9 pile
Flordecke (f)
pelo (m)
poil (m)
pelo (m)

10 pile fabric, automotive
Velours (n) für KFZ-Bereich
tela (f) con pelo para automóviles
tissu (m) à poil pour automobiles
tessuto a pelo (m) per automobili

11 pile fabric manufacturers
Flordeckenhersteller (m pl), Plüsch-Hersteller (m pl)
fabricantes (m pl) de tela de pelo
fabricants (m pl) de tissu à poils
produttori (m pl) di tessuto a pelo

12 pilling
Pilling (n)
formación de bolas
boulochage (m)
pilling/di pilling formazione (f)

13 pirn
Rolle (f), Hülse (f)
carrete (m)
canette (f)
cannetta (f), rocchetto (m), spola (f)

14 plain knitting
Glattstrickerei (f)
puntos (m pl) sencillos
tricotage (m) uni
lavorazione (f) a punto legaccio

15 plain weave
Grundbindung (f), (Tuchbinding, Taftbinding, Leinwandbinding)
tejido (m) sencillo
armure (f) unie
armatura tela

16 plastic
Plastik (n), Kunststoff (m)
plástico (m), material (m) sintético
plastique (m), matière (f) plastique
materia (f) plastica

17 plastic
plastisch (adj)
plástico
plastique
plastico

18 plastic cones and tubes
Kunstoffkegel und -zylinder, -conen, -hülsen
conos (m pl) y tubos (m pl) plásticos
cônes (m pl) et tubes (m pl) en plastique
rocchetti (m pl) e fusi di plastica

1 pleat
Falte (f)
pliegue (m)
plissé (m), pli (m)
piega (f)

2 pleat (f)
plissieren, falten
plisar
plisser
piegare

3 ply
Lage (f)
capa (f)
pli (m)
ritorto

4 pneumatic
pneumatisch (adj)
neumático
pneumatique
pneumatico

5 polyamide
Polyamid (n)
poliamida (f)
polyamide (m)
poliammide (f)

6 polyester
Polyester (n)
poliéster (m)
polyester (m)
poliestere (m)

7 polyester staple
Polyesterstapel (m)
hebra (f) poliéster
fibre (f) de polyester
fiocca (f) poliestere

8 polypropylene
Polypropylen (n)
polipropileno (m)
polypropylène (m)
polipropilene (m)

9 polypropylene – bags and films
Polypropylen (n) – Säcke (m pl) und Folien (f pl)
sacos (m pl) y membranas (f pl) de polipropileno
sacs (m pl) et films (m pl) en polypropylène
polipropilene (m) – sacchetti e pellicole

10 polypropylene – box packs
Polypropylen (n) – Box-Packungen (f pl)
cajas (f pl) de polipropileno
boîtes (f pl) en polypropylène
polipropilene (m) – scatole e scatoloni

11 polypropylene – lubricants
Polypropylen (n) – Gleitmittel (n pl)
lubrificantes (m pl) de polipropileno
lubrifiants (m pl) pour polypropylène
polipropilene (m) – lubrificanti

12 polypropylene – recyclers
Polypropylen (n) – Wiederverwerter (m pl)
recirculadores (m pl) de polipropileno
recyclage (m) de polypropylène
polipropilene (m) – riciclatrici

13 polypropylene – strapping
Polypropylen (n) – Gurtstoff (m)
correaje (m) de polipropileno
liens (m pl) en polypropylène
polipropilene (m) – reggette

14 polypropylene – twines
Polypropylen (n) – ßindfaden (m pl)
guitas (f pl) de polipropileno
ficelle (f) en polypropylène
polipropilene (m) – spaghi di polipropilene

15 polythene
Polyäthylen (n)
polietileno (m)
polyéthylène (m)
politene (m)

16 polythene bags
Polyäthylensacke (m pl)
sacos (m pl) de polietileno
sacs (m pl) en polyéthylène
sacchetti (m pl) di politene

17 polythene packaging
Polyäthylenverpackung (f)
embalaje (m) de polietileno
emballage (m) en polyéthylène
politene (m) per imballagio

18 polythene – shrink film
Polyethylen (n) – Schrumpffolie (f)
pelicular encogible de polietileno
film (m) retractable en polythène
politene (m) – pellicola restringibile

1 polythene – squares
Polyethylen (n) – Bögen (m pl)
cuadros (m pl) de polietileno
carrés (m pl en polythène
politene (m) – quadrati

2 porous
porös
poroso
poreux
poroso

3 preparation
Vorbereitung (f)
preparación (f)
préparation (f)
preparato (m), preparazione (f)

4 presentation aids
Präsentationsmittel (n pl)
auxiliares (m pl) de presentación
aides (f pl) à la présentation
strumenti (m pl) di aiuto per la presentazione

5 press
Presse (f)
prensa (f)
presse (f)
pressa (f)

6 pressure
Druck (m)
presión (f)
pression (f)
pressione (f)

7 pressure dyeing
Druckfärben (n)
teñir (m) a presión
teinture (f) sous pression
tintura (f) a pressione

8 print
Druck (m)
estampado
empreinte (f)
stampa (f)

9 printed cloth
bedruckter Stoff (m)
tela (f) estampada
tissu (m) imprimé
tessuto (m) stampato

10 printers and finishers
Drucker und Veredler (m pl)
estampadores (m pl) y rematadores (m pl)
imprimeurs (m pl) et apprêteurs (m pl)
stampatori e finitori (m pl)

11 printers – apparel
Drucker (m pl) – Bekleidung (f)
ropa (f) de estampadores
imprimeurs (m pl) – vêtements
stampatori-vestiario (m)/abbigliamento (m)

12 printers – furnishing
Drucker (n pl) – Möbel (f)
equipos (m pl) estampadores
imprimeurs (m pl) – ameublement
stampatori (m pl) – arredamento

13 printers – heat transfer
Drucker (n pl) – Hitzetransfer (m)
estampadores -transmissión térmica
imprimeurs (m pl) – transmission thermique
stampatori (m pl) – trasferimento di calore

14 printers textile
Textildrucker (m pl)
estampadores (m pl) de tejidos
imprimeurs (m pl) textiles
stampatori (m pl) tessili

15 printers textile [thickness]
Textildrucker (m pl) [Stärke]
estampadores (m pl) de tejidos [espesor]
imprimeurs (m pl) textiles [épaisseur]
stampatori (m pl) tessili [spessore]

16 process
Prozeß (m)
proceso (m)
processus (m)
processo (m)

17 produce (farm)
Produkt (n)
producto (m)
produits (m pl)
prodotto (m) agricolo

18 produce (v)
produzieren, fabrizieren
producir, fabricar
produire, fabriquer
produrre

19 producer
Produzent (m)
fabricante (m)
producteur (m) fabricant
produttore (m)

1 production
Produktion (f)
producción (f)
production (f)
produzione (f)

2 productivity
Produktivität (f)
productividad (f)
productivité (f)
produttività (f)

3 protective coatings
Schutzschicht (f)
capas (f pl) de protección
couches (f pl) de protection
rivestimenti (m pl) protettivi

4 Prussian blue
Berlin Blau (n)
azul de Prusia
bleu de Berlin
blu di Prussia

5 publications, textile
Publikationstextil (m)
publicaciones (f pl) de tejidos
publications (f pl) textiles
pubblicazioni (f pl) tessili

6 public conditioning house
öffentliche Konditionieranlage (f)
casa (f) pública de acondicionamiento
condition (f) publique textile
stablimento (m) pubblico di condizionatura

7 pucker
Faltenbausch (m)
frunce (m)
fronce (f)
grinza, crespa, ruga (f)

8 pucker (v)
zusammenziehen, falten
fruncir
froncer
raggrinzare, increspare

9 pullers
Verzieher (m pl)
tiradores (m pl)
arracheurs (m pl)
tiratori (m pl)

Q

10 quality
Qualität (f)
calidad (f)
qualité (f)
qualità (f)

11 quality control
Qualitätskontrolle (f)
control (m) de calidad
contrôle (m) de qualité
controllo (m) di qualità

12 quantity
Menge (f)
cantidad (f)
quantité (f)
quantità (f)

R

13 rag
Lumpen (m)
trapo (m)
chiffon (m)
straccio (m)

14 rag pulling
Lumpenreißen (n)
deshilachado de trapos
effilochage (m) des chiffons
sfilacciatura (f) di stracci

15 rags
Lumpen (m pl)
trapos (m pl)
chiffons (m pl)
stracci (m pl)

16 rags – wiper
Lumpen (m pl) – Wischtuch (m)
trapos (m pl) para limpiar
chiffons (m pl) à essuyer
stracci (m pl) – per pulire

1 raising
Tuchrauhen (n)
perchado
lainage (m)
garzatura (f)

2 raised finish
Rauhappretur (f)
remate (m) perchado
gratté
rifinitura (f) garzata

3 ramie
Ramie (f)
ramio (m)
ramie (f)
ramiè (m)

4 rapier
Lanze (f), Lanzenwebstuhl (m) [Webstuhl], Greifer (m)
espadín (m)
lance (f) rigide
pinza (f), uncinetto (m)

5 raw
roh, unbehandelt (adj)
crudo
cru
grezzo

6 raw materials
Rohmaterialien (n pl)
materias primas
matières (f pl) brutes
materiali (m pl) grezzi

7 raw silk
Rohseide (f)
seda (f) cruda
soie (f) crue
seta (f) grezza

8 rayon
Reyon (n)
rayón
rayonne (f)
raion (m)

9 recombers and gillers
Nachkämmer (m pl) und Hecheler (m pl)
repeinadores (m pl)
repeigneurs (m pl) et gillers (m pl)
ripettinatrici (f pl) e pettinatrici

10 refractometers
Refraktometer (m)
refractómetros (m pl)
réfractomètres (m pl)
rifrattometri (m pl)

11 research – textile industry
Forschung (f) – Textilforschung (f)
investigaciones (f pl) de la industria textil
recherche (f) textile
ricerca (f) – industria tessile

12 reaching in
Zureichen (n), Eintichen (n)
alcanzar (m)
donner les fils
porgitrice (f) di fili

13 recombing
Nachkämmen (n)
peinar (m) de nuevo
double peignage
ripettinatura (f)

14 reed [weaving]
Rietblatt (n)
peine (m)
peigne (m) miseur
pettine (m), numero di fili di ordito per pollice [tessitura]

15 reed mark [weaving]
Rietstreifen (m)
marca (f) de peine
défaut (m) de peigne
segno (m) di dente del pettine/rigature del pettine (f pl)

16 reed width [weaving]
Rietweite (f)
anchura (f) de peine
largeur (f) de peigne
altezzo in pettine (f)

17 reeling
Haspeln (n)
devanar (m)
dévidage (m)
incannatura (f)/aspatura (f)

18 regain
Reprise (f)
recuperacíon (f)
reprise (f)
ripresa (f), tolleranza di umidità

1 regenerated
wiederaufbereitet
regenerado
régeneré
rigenerato

2 registered
eingetragen
certificado
rapporté/désposé
registrato

3 regular
gleichmäßig
regular
régulier
regolare

4 regulation
Regelung (f)
regulación (f)
regolazione (f)

5 relaxing
Entspannen (n)
aflojamiento (m)
relaxation (f)
rilassare

6 removal
Entfernung (f), Beseitigung (f)
retirado (m)
enlèvement (m)
rimozione (f)

7 repp
Rips (m)
tela (f) de cordoncillo
reps (m)
reps (m)

8 residue
Rückstand (m), Überrest (m)
residuo (m)
résidu (m)
residuo (m)

9 resin
Harz (n)
resina (f)
résine (f)
resina (f)

10 reverse twist
Linksdrehung (f)
torsión (f) al revés
torsion (f) en S
torsione (f) inversa

11 rewinding
Rückspulen (n), Umspulen (n)
redevanar (m)
renvidage (m)
stracannatura (f)/riavvolgimento (m)

12 ribbon
Band (n)
cinta (f)
ruban (m)
nastro (m)

13 ring frame
Ringspinnmaschine (f)
tejido de anillo
continu (m) à anneau
filatoio (m) ad anelli

14 ring spinning
Ringspinnen (n)
hilatura de anillo
filature (f) à anneau
filatura (f) ad anelli

15 ring twisting
Ringzwirnen (n)
tursión de anillo
retordage (m) à anneau
ritorcitura (f) ad anelli

16 roll [cloth]
Rolle (f)
rollo (m)
enrouler
rullo (m), cilindro (m)

17 roller
Walze (f)
tambor (m)
rouleau (m)
rullo (m)

18 roller coverings
Walzenbug (m)
cubiertas (f pl) de tambores
revêtements (m pl) du rouleau
coperture (f pl) per rulli/rivestimenti del rullo

19 rollers
Walzen (f pl)
tambores (m pl)
rouleaux (m pl)
rulli (m pl)

1 rollers – rubber
Walzen (f pl) – Gummi (n)
arrolladores (m pl) de goma
cylindres (m pl) en caoutchouc
rulli (m pl) di gomma

2 rolling boards
Walzbretter (m pl)
tablas (f pl) para arrollarse
tableaux (m pl) roulants
cartoni (m pl) da laminazione

3 rope
Seil (n), Strang (m)
cuerda (f)
boyau (m)
corda (f), fune (f)

4 rope scouring
Strangwaschen (n)
cuerda (f) desengrasada
lavage (m) en boyau
lavaggio (m) della corda

5 ropes and twines
Stränge (m pl) und Bindfäden (m pl)
cuerdas (f pl) y guitas (f pl)
corderie (f)
corde (f pl) e spaghi (m pl)

6 rotproof
fäulnisecht
protegido contra putrefacción
résistant à la pourriture
anti putrefazione/resistente alla putrefazione

7 roving
Vorgarn (m)
retorcer (m)
mèche (f)
lucignolo (m), stoppino (m)

8 rubbing
Reiben (n)
frotar (m)
frottement (m) abrasif
frizione (f)/frottatura (f)

9 rug – cashmere, wool and blends
Teppich (m) – aus Kaschmir, Wolle und Mischungen
alfombrillas (f pl) de cachemir, lana y mezclas
tapis (m pl) cachemire, laine et mélanges
tappeto (m) – cashmere, lana e misti

10 rug manufacturers
Teppichhersteller (m pl)
fabricantes (m pl) de alfombrillas
fabricants (m pl) de tapis
produttori (m pl) di tappeti

S

11 sack sewing threads
Sacknähzwirne (m pl)
hilos (m pl) para coser sacos
fils (m pl) à coudre les sacs
filati (m pl) cucirini per sacchi

12 safety products
Sicherheitsprodukt (n)
productos (m pl) de seguridad
produits (m pl) de sécurité
prodotti (m pl) di sicurezza

13 salary
Gehalt (n)
salario (m)
traitement (m), salaire (m)
stipendio (m)

14 sale
Verkauf (m)
venta (f)
vente (f)
vendita (f)

15 salt
Salz (n)
sal (f)
sel (m)
sale (m)

16 sample
Muster (n), Probe (f)
muestra (f)
échantillon (m)
campione (m)

17 sample cutters
Musterschneider (m pl)
cortadores (m pl) de muestras
découpeuses (f pl) d'échantillons
(f pl) taglierine per campioni

18 satin
Satin (m)
raso (m)
satin (m)
satin (m)

1 scarves
Schals (m pl)
bufandas (f pl)
écharpes (f pl), foulards (m pl)
sciarpe (f pl)

2 scorch (v)
versengen
chamuscar
brûler
bruciare

3 scouring
Reinigen (n)
desengrasar (m)
lavage (m) à fond
lavaggio (m)

4 scribbling
Grobkrempeln (n), Vorkrempeln (n)
cardadura (f)
précardage (m)
cardatura (f)

5 selection
Auswahl (f)
selección (f)
sélection (f)
selezione (f)

6 selvedge
Webrand (m), Webkante (f)
orillo (m)
lisière (f)
cimosa (f)

7 semi-automatic
halbautomatisch
semi automático
semi-automatique
semi-automatico

8 semi worsted spinning
Semikammgarnspinnen (n)/Halbkammgaonspinnen (m)
hilar (m) de semi estambre
filature (f) mi-peignée
filatura (f) di semi pettinati

9 sewing thread
Nähzwirn (m)
hilo (m) de coser
fil (m) à coudre
filo da cucire

10 sewing thread manufacturers
Nähfadenhersteller
fabricantes (m pl) de hilos de coser
fabricants (m pl) de fil à coudre
produttori (m pl) di filo da cucire

11 shade
Schattierung (f)
matiz (m)
nuance (f)
tinta (f)

12 shade card
Schattierungskarte (f), Farbkarte (f)
carta (f) de matices
carte (f) de nuances
cartella (f) colori

13 shading
Abtönen (n)
matizar
effet (m) de dégradé
sfumatura (f)

14 shearing [cloth]
Stoffscheren (n)
tundición (f)
tondage (m)
cimatura (f) [tessuto]

15 shear [sheep]
scheren
esquilar
tondre
tosare [pecore]

16 sheets – centre fold
Laken – Mittelfalten
hojas (f pl) del centro
draps (m pl) à pli central
lenzuoli (m pl) – piegatura al centro/piegatura centrale

17 shoddy
Shoddy (n); wertlos
paño (m) burdo
tissu (m) de renaissance
lana (f) rigenerata

18 shoddy manufacturers
Shoddyhersteller (m pl)
fabricantes (m pl) de paño burdo
fabricants (m pl) de tissu de renaissance
produttori (m pl) di lana rigenerata

1 shrinkage
Eingang (m), Schrumpfung (f), Einlaufen (m)
encogimiento (m)
retrait (m)
restringimento (m)/accorciamento (m)

2 shrinkproof
schrumpffest
a prueba de encogimiento
non-rétrécissant, solide au retrait
irrestringibile

3 shrinking
Schrumpfen (n)
ecogimiento
rétrécissement (m)
restringimento (m)

4 shrinkproof treatment
Schrumpfschutzbehandlung (f)
tratamiento (m) contra el encogido
apprêt (m) antiretrait
trattamento (m) anti-accorciamento

5 shuttle
Schiffchen (n), Schützen (m)
lanzadera (f)
navette (f)
navetta (f), spola (f)

6 shuttleless loom
schützenloser Webstuhl (m)
telar (m) sin lanzadera
métier (m) sans navette
telaio (m) senza navette

7 silk
Seide (f)
seda (f)
soie (f)
seta (f)

8 size
Größe (f)
aprestar (m)
encoller (v)/dimension (f)
dimensione (f)

9 sizers
Schlichter (m pl)
aprestadores (m pl)
encolleuses (f pl)
calibratrici (f pl)

10 skein
Garnsträhne (f), Strang (m)
madeja (f)
écheveau (m) de fil
matassina (f)

11 sley [loom]
Lade (f)
peine (m)
battant (m)
battente (m) [telaio]

12 slipes
Gerberwolle (f)
lana (f) tirada de piel
laine (f) d'enchaussenage
lana (f) calcinata

13 slipes – New Zealand greasy
Hautwolle (f) – Neuseeland – schmierig
lana (f) grasienta de piel de Nueva Zelanda
laine (f) d'enchaussenage en suint de la Nouvelle-Zélande
lana (f) calcinata – grassa – Nuova Zelanda

14 slipes – New Zealand scoured
Hautwolle (f) – Neuseeland – gewaschen
lana (f) desengrasada de piel de Nueva Zelanda
laine (f) d'enchaussenage lavée à fond de la Nouvelle-Zélande
lana (f) calcinata – lavata – Nuova Zelanda

15 slippage
Verschiebung (f)
resbalamiento (m)
glissement (m)
scorrimento (m)/slittamento

16 sliver
Kardenbänder (n pl)
mecha (f)
ruban (m) de fibres
nastro (m) di carda/pettinato

17 sliver cans
Kardenbandbüchen (f pl)
latas (m pl) para mechas
pots (m pl) de ruban
vasi portanastri

1 slivers
Spinnbänder (m pl)
tiritas (f pl)
rubans (m pl) de fibres
nastri (m pl) di carda

2 sliverers
Kardenbinder (m pl)
fabricantes (m pl) de mechas
réunisseuses (f pl) de ruban
riunitrici (f pl) dei nastri

3 slub
Fadenverdickung (f), Flamme (f)
mechón (m)
mèche (f)
ingrossamento (m) del filato/bottone (m)

4 slub catching
Fadenreinigen (m)
cogida (f) de mechones
épuration (f) de fil
pulitore (m) dei fili

5 slubbing
Luntenspinnen (n)
torcer (m) lana
filage (m) en gros
filatura (f) in grosso/stoppino (m)

6 slubbing dyed
gefärbte Lunten (f pl)
mechones (m pl) teñidos
teint sur ruban peigné
filatura (f) in grosso tinta

7 slubs
Fadenverdickungen (f pl), Flammen (f pl)
mechones (m pl)
fils (m pl) en gros
ingrossamenti (m pl) del filato

8 smooth
glatt
suave
lisse
liscio

9 snagging
Zieheranfälligkeit (f), Fadenziehen (m)
enganchar (m)
formation (f) d'accroc
smagliatura (f)

10 snarling
Schlingenbildung (f)
enredar (m)
vrille (f)
ricci (m pl)

11 soap
Seife (f)
jabón (m)
savon (m)
sapone (m)

12 soaps
Seifen (f pl)
jabones (m pl)
savons (m pl)
saponi (m pl)

13 soaps – textile
Seifen (f pl) – textil
jabones (m pl) para textiles
savons (m pl) textiles
saponi (m pl) – tessili

14 sodium sulphate
Natriumsulfat (n), Glaubersalz (n)
sulfato (m) de sodio
sulfate (m) neutre de sodium
solfato (m) di sodio

15 softener
Weichmacher (m), Enthärter (m)
suavizante (m)
agent (m) adoucissant
ammorbidente (m)

16 softeners
Weichmacher (m pl), Enthärter (m pl)
suavizantes (m pl)
assouplissants (m pl)
ammorbidenti (m pl)

17 soft toy trade fillings
Stofftierfüllungen (f pl)
rellenos (m pl) para la industria de juguetes blandos
rembourrages (m pl) pour l'industrie des peluches
imbottitura (f) – per giocattoli (peluches)

18 soft twist
weiche Drehung (f)
torsión (f) suave
torsion (f) floche
torsione (f) leggera

1 solution
Lösung (f)
solución (f)
solution (f)
soluzionc (f)

2 solvent
Lösungsmittel (n)
disolvente (m)
solvant (m)
solvente (m)

3 solvents
Lösungsmittel (n pl)
disolventes (m pl)
solvants (m pl)
solventi (m pl)

4 sorting
Sortieren (n)
clasificar (m)
tri (m) classification (f)
cernita (f)

5 space dyed
in Abständen gefärbt
teñido a trechos
teint par espacement
tintura (f) di sezioni

6 speciality fibres
Spezialfasern (f pl)
fibras (f pl) especiales
fibres spéciales
fibre (f pl) speciali

7 specification
Beschreibung (f) [technische]
especificación (f)
spécification (f)
specifica (f)/specificazione

8 specific gravity
spezifisches Gewicht (n)
densidad (f)
poids (m) volumique
peso (m) specifico

9 spin (v)
spinnen, drehen
hilar (m)
filer
filare

10 spindle
Spindel (f)
huso (m)
broche (f)
fuso (m)

11 spin finishers
Spinnappretierer (m pl)
rematadores (m pl) de hilatura
finisseuses (f pl) de filage
finitori (m pl) di filato

12 spinning
Spinnen (n)
hilatura
filature (f)
filatura (f), filato (m)

13 spool
Spule (f)
carrete (m)
bobine (f) de filé
cabina (f) rocca (f) rocchetto (m)

14 spot
Fleck (m)
punto (m)
tâche (f)
macchia (f)

15 spray
Spray (n)
pulverizar
pulvériser (v)
spray (m)

16 springs
Feder (f)
muelles (m pl)
ressorts (m pl)
molle (f pl)

17 standard
Standard (m), Norm (f)
norma (f)
norme (f)
standard (agg), normale

18 standard time
Normalzeit (f)
tiempo (m) normal
temps (m) standard
temps (m) standard
ora (f) legale

1 staple fibre
Kurzfaser (f), Stapelfaser (f)
fibra (f) clasificada
fibre (f) coupée
fibra (f) fiocco

2 staple fibre-acrylic
Kurzfaser / Acrylkurzfaser (f)
fibra (f) acrílica clasificada
fibre (f) acrylique coupée
fibra (f) fiocco acrilica

3 staple fibre-nylon
Kurzfaser / Nylonkurzfaser (f)
fibra (f) clasificada de nilón
fibre (f) de nylon coupé
fibra (f) fiocco nailon

4 staple fibre-polyester
Kurzfaser / Polyesterkurzfaser (f)
fibra clasificada de poliéster
fibre (f) de polyester coupée
fibra (f) fiocco poliestere

5 staple length
Stapellänge (f), Faserlänge (f)
largura (f) clasificada
longeur (f) de fibres
lunghezza (f) di fiocca

6 static eliminator
Antistatikvorrichtung (f)
eliminador (F) de electricidad estática
éliminateur (m) d'électrícité statique
eliminatore (m) di cariche elettrostatiche

7 steam
Dampf (m)
vapor (m)
vapeur (m)
vapore (m)

8 steam conversion systems
Dampfumwandlungsysteme (n pl)
sistemas (m pl) de conversión de vapor
systèmes (m pl) de conversion de vapeur
sistemi (m pl) di conversione del vapore

9 steeplejacks
Stapler (m pl) / Arbeiter für Reparatur/Wartung hochgelegener Maschinen/Maschinenteile
reparadores (m pl) de espiras
réparateurs (m pl) de cheminées d'usine
banco (m) a fusi in finissimo a guglia

10 stock
Bestand (m)
fibras (f pl) sueltas no elaboradas
matière (f) brute
stock (m), magazzino (m)

11 stock dyeing
Flockefärben (n)
teñido (m) de fibras sueltas no elaboradas
teinture (f) en bourre
tintura in fiocco

12 stockinette
Baumwolltrikot (n)
elástica (f)
tricot (m)
tessuto (m) a maglia

13 stocktaking
Bestandsaufnahme (f), Inventur (f)
inventario (m)
inventaire (m)
inventario (m)

14 stoles
Stolas (f pl)
estolas (f pl)
étoles (f pl)
stole (f pl)

15 stop motion [loom]
Abstellvorrichtung (f)
parada (f) del movimiento
casse-ruban (m)
arresto (m) automatico [telaio]

16 storage
Lagerung (f)
almacenaje (m)
stockage (m)
magazzinaggio (m)

17 storage – all fibres
Lagerung (f) – alle Fasern
almacenaje (m) de toda clase de fibra
emmagasinage (m) – toutes fibres
magazzinaggio (m) – tutte le fibre

18 strapping
Gurtstoff (m), Angurten (n)
correaje (m)
liens (m pl)
cinghie (f)

1 strength
Stärke (f)
fuerza (f)
résistance (f)
resistenza (f) forza (f) potenza (f)

2 stretch
Streckung (f), Dehnung (f)
estirar (v)
extensibilité (f) élastique
allungamento (m) estensione (f)
stiro (m)

3 stripe
Streifen (m)
raya (f)
rayure (f)
striscia (f), rigatura di colore, segno
(m) del pettine

4 stripiness
Streifigkeit (f)
ilusión (f) de rayas
aspect (m) rayé
formazione (f) di righe

5 S-twist
S-Drehung (f)
torcedura (f) S
tordu S
torsione (f) S

6 substitute
Ersatzstoff (m)
substituto (m)
remplacement (m)
sostituto (m)

7 surgical dressings
Verbandsmaterial (n)
vendas (f pl) quirúrgicas
pansements (m pl)
garze (f pl) chirurgiche

8 suture threads
Wundnähfäden (m pl)
hilos (m pl) de sutura
fils (m pl) pour sutures
punti (m pl) di sutura/fili di sutura

9 swatch
Musterstreifen (m), Musterabschnitt
(m), Liasse (f)
muestra (f) de tela
coupon (m) d'échantillon
campioncino (m)

10 synthetic
synthetisch
sintético
synthétique
sintetico

T

11 take up motion [loom]
Aufwickelvorrichtung (f)
recogido (m) de urdimbre
régulateur (m) d'enroulement du tissu
meccanismo (m) di avvolgimento
[telaio]/dispositive (m) di
avvolgimento

12 tape condenser
Riemenflorteiler (m)
condensador (m) de cintas
diviseur (m) à lanières
divisore (m) a cinghiette

13 tapes
Bänder (n pl)
cintas (f pl)
bandes (f pl)
nastri (m pl)

14 tapes – adhesive
Bänder (n pl) – Klebe-
Bänder
cintas (f pl) adhesivos
bandes (f pl) adhésives
nastri (m pl) – adesivi

15 tapes – pressure sensitive
Bänder (n pl) – druckempfindliche
Bänder
cintas (f pl) autoadhesivas
bandes (f pl) autoadhésives
nastri (m pl) – sensibili alla
pressione

16 tapes – polypropylene
Bänder (n pl) – Polypropylen-
Bänder
cintas (f pl) de polipropileno
bandes (f pl) en polypropylène
nastri (m pl) – polipropilene

17 tapes – spindle
Bänder (n pl) – Spindel-
Bänder
cintas (f pl) para husos
bandes (f pl) pour broches
nastri (m pl) – fuso

1 tapestry
Wandteppich (m), Wandbehang (m), Möbelstoff (m)
tapicería (f)
tapisserie (f)
tappezzeria (f)

2 tappet loom
Exzenterstuhl (m)
telar (m) de levas
métier (m) à came
telaio (m) – a accentrici

3 tear
Riß (m)
rasgón (m)
déchirer (v)
strappo (m)

4 teasle
Weberdistel (f)
cardencha (f)
chardon (m) à gratter
garza (f)

5 teasle gig
Weberdistelrauhmaschine (f)
máquina (f) de cardenchas
laineuse (f) à chardons
garzatrice

6 teasles
Weberdisteln (f pl)
cardenchas (f pl)
chardons (m pl)
garze (f pl)

7 teasles [wire]
Rauhkarden [Ruten] (f pl)
cardas (f pl) de alambre
cardes (f pl) [fil métallique]
carde (f pl) [spazzola metallica]

8 technical
technisch
técnico
technique
tecnico

9 temperature
Temperatur (f)
temperatura (f)
température (f)
temperatura (f)

10 temperature control
Temperaturkontrolle (f)
regulación (f) de temperatura
réglage (m) de température
controllo (m) della temperatura

11 temple [loom]
Spannstab (m), Breithatter (m)
vara (f)
métier (m) à templet
tempiale (agg/m) [telaio]

12 tensile strength
Zugfestigkeit (f)
resistencia (f) tensil
résistance (f) à la traction
resistenza (f) alla trazione

13 tentering [drying]
Spannrahmentrocknung (f)
tender (m)
traitement (m) sur rame
trattamento in rameuse

14 test
Test (m); testen (v)
prueba (f)
essai (m)
prova (f), test (m)

15 testing
Testen (n)
probar (m)
contrôle (m)
prova (f), esame (m)

16 textile
Textil (n), Gewebe (n)
tejido (m)
textile (adj)
tessile (agg)

17 textile accessories
Textilzubehör (n) (kein Plural – no plural)
accesorios (m pl) de tejidos
accessoires (m pl) textiles
accessori (m pl) tessili

18 texture
Struktur (f)
textura (f)
texture (f)
tessuto (m)/tessitura (f)

1 three ply
Dreifach-
de tres capas
à trois brins
ritorto (m) di tre capi

2 throwsters
Seidenzwirner (m)/Seidenspinner (m)
torcedores (m pl) de seda
mouliniers (m pl)
torcitori di seta

3 tints, fugitive
zarte Farben (f pl), flüchtige Farben (f pl), Signierfarben (f pl)
matices (m pl) fugitivos
colorants (m pl) fugaces
colori (m pl) non saturi

4 top
Kammzug (m)
carrete (m) mecha
ruban (m)
nastro (m)/top

5 top dyed
gefärbter Kammzug (m)
teñido (m) en mecha
teint en rubans peignés
nastro (m) tinto

6 top maker
Kammzugmacher (m)
fabricante (m) de mechas
fabricant (m) de rubans peignés
nastro (m) – fabbricante

7 tops – acrylic
Kammzüge (m pl) – Acrylkammzug (m)
mechas (f pl) de hilo acrílico
rubans (m pl) acryliques
nastri (m pl) – acrilici

8 tops – acrylic-coloured
Kammzüge (m pl) – Acryl gefärbt
mechas (f pl) acrílicas de colores
rubans (m pl) acryliques teints
nastri (m pl) – colorati – acrilici

9 tops – agent
Kammzüge (m pl) – Verteter für Kammzüge
agente (m) de mechas
agents (m pl) pour rubans
nastri (m pl) – agente

10 tops – alpaca
Kammzüge (m pl) – Alpakakammzug (m)
mechas (f pl) de pelo de alpaca
rubans (m pl) d'alpaga
nastri (m pl) – alpaca

11 tops and slivers – fancy effects
Kammzüge (m pl) – Spinnbänder – Fantasieeffekt
mechas y cintas – adornos
rubans (m pl) et mèches (f pl) – effets fantaisie
nastri (m pl) e nastri di carda – effetti fantasia

12 tops – angora
Kammzüge (m pl) – Angorakammzug (m)
mechas (f pl) de pelo de angora
rubans (m pl) d'angora
nastri (m pl) – angora

13 tops – blended
Kammzüge (m pl) – gemischter Kammzug (m)
mechas (f pl) de hilos mezcladas
rubans (m pl) mixtes
nastri (m pl) – misti

14 tops – British wool
Kammzüge (m pl) – britischer Wollkammzug (m)
mechas (f pl) de lana británica
rubans (m pl) de laine britannique
nastri (m pl) – lana inglese

15 tops – broken
Kammzüge (m pl) – unregelmäßiger Kammzug (m)
mechas (f pl) de hilos rotos
rubans (m pl) cassés
nastri (m pl) – sfilacciati/rasati

16 tops – camel hair
Kammzüge (m pl) – Kamelhaar-kammzug (m)
mechas (f pl) de pelo de camello
rubans (m pl) de poil de chameau
nastri (m pl) – pelo di cammello

1 tops – cape – worsted
Kammzüge (m pl) – Capekammgarn (n)
mechas (f pl) de estambre
rubans (m pl) pour capes peignées
nastri (m pl) – mantella – pettinati

2 tops – cashmere
Kammzüge (m pl) – Kashmirkammzug (m)
mechas (f pl) de cachemir
rubans (m pl) de cachemire
nastri (m pl) di cashmere

3 tops – coloured
Kammzüge (m pl) – färbiger Kammzug (m)
mechas (f pl) de hilos de colores
rubans (m pl) colorés
nastri (m pl) – colorati

4 tops – crossbred
Kammzüge (m pl) – Crossbred-kammzug (m)
mechas (f pl) de lanas mezcla
rubans (m pl) croisés
nastri (m pl) – lana incrociata

5 tops – crossbred hosiery
Kammzüge (m pl) – Crossbred-strumpfwaren (f pl)
mechas (f pl) mezclas calcetería
rubans (m pl) de bonneterie croisés
nastri (m pl) lana incrociata – maglieria

6 tops – crossbred lustre
Kammzüge (m pl) – Crossbredglanz (m)
mechas (f pl) mezcla y brillantes
rubans (m pl) de lustrine croisés
nastri (m pl) – lane incrociate – brillanti

7 tops – English wool
Kammzüge (m pl) – englischer Wollkammzug (m)
mechas (f pl) de lana inglesa
rubans (m pl) de laine anglaise
nastri (m pl) – lana inglese

8 tops – exporters
Kammzüge (m pl) – Kammzug-exporteure (m pl)
mechas (f pl) – exportadores
exportateurs (m pl) de rubans peignés
nastri (m pl) – esportatori

9 tops – goat hair
Kammzüge (m pl) – Ziegenhaar-kammzug (m)
mechas (f pl) de pelo de cabra
rubans (m pl) de poil de chèvre
nastri (m pl) – pelo di capra

10 tops – hand knitting
Kammzüge (m pl) – für Handstrick
mechas (f pl) para hacer puntos a mano
rubans (m pl) pour tricot à la main
nastri (m pl) lavoratori a mano

11 tops – lambswool
Kammzüge (m pl) – Lammwoll-kammzug (m)
mechas (f pl) de lana de cordero
rubans (m pl) de laine d'agneau
nastri (m pl) – lambswool

12 tops – llama
Kammzüge (m pl) – Lamakammzug (m)
mechas (f pl) de pelo de llama
rubans (m pl) de lama
nastri (m pl) – lama

13 tops – manmade
Kammzüge (m pl) – Kunstfaser-kammzug (m)
mechas (f pl) de hilos sintéticos
rubans (m pl) artificiels
nastri (m pl) – artificiali

14 tops – merino
Kammzüge (m pl) – Merino-schafkammzug (m)
mechas (f pl) de lana merina
rubans (m pl) de mérinos
nastri (m pl) – di pecora merino

15 tops – mohair
Kammzüge (m pl) – Mohairkammzug (m)
mechas (f pl) de pelo de angora
rubans (m pl) de mohair
nastri (m pl) – mohair

16 tops – polyamide
Kammzüge (m pl) – Polyamid-kammzug (m)
mechas (f pl) de hilos de poliámida
rubans (m pl) de polyamide
nastri (m pl) – poliammide

1 tops – polyester
Kammzüge (m pl) – Polyester-kammzug (m)
mechas (f pl) de hilos de poliéster
rubans (m pl) de polyester
nastri (m pl) – poliestere

2 tops – silk
Kammzüge (m pl) – Seidenkammzug (m)
mechas (f pl) de hilo de seda
rubans (m pl) de soie
nastri (m pl) – seta

3 tops – synthetic
Kammzüge (m pl) – Synthetik-Kammeug (m)
mechas (f pl) de hilos sintéticos
rubans (m pl) synthétiques
nastri (m pl) – sintetici

4 tops – wool
Kammzüge (m pl) – Wollkammzug (m)
mechas (f pl) de lana
rubans (m pl) de laine
nastri (m pl) – lana

5 tops – yak hair
Kammzüge (m pl) – Yak-Kammzug (m)
mechas (f pl) de pelo de yak
rubans (m pl) de poil de yak
nastri (m pl) – pelo di yak

6 tow
Werg (m), Spinnband (n)
estopa (f)
câble (m) de filaments
stoppa (f)

7 tow – acrylic
Acrylspinnband (n)
estopa (f) acrīlica
câble (m) de filaments acrylique
stoppa – acrilica

8 tow dyed
gefärbtes Spinnband (n)
estopa (f) teñida
teint sur câble
stoppa (f) – tinta

9 tow – nylon
Nylonspinnband (n)
estopa (f) de nilón
câble (m) de filaments nylon
stoppa (f) – di nailon

10 tow – polyester
Polyesterspinnband (n)
estopa (f) de polieśster
câble (m) de filaments polyester
stoppa (f) di poliestere

11 transaction
Transaktion (f)
transacción (f)
opération (f) commerciale
transazione (f)

12 translation services
Übersetzungsdienste (f pl)
servicios (m pl) de traduccion
service (m) de traduction
servizi (m pl) di traduzione

13 transport
Transport (m)
transporte (m)
transport (m)
trasporto (m)

14 transport and haulage contractors
Transport und Speditionsunternehmer (m pl)
contratistas (m pl) de transporte y acarreos
entrepreneurs (m pl) de transports
imprenditori (m pl) di trasporti

15 travel rugs
Reisedecken (f pl)
mantas (f pl) de viaje
couvertures (f pl) de voyage
plaid (m)/coperte (f pl) da viaggio

16 trial
Versuch (m)
ensayo (m)/prueba
essai (m)
prova (f)

17 trucks and trolleys
Karren und Schienenwagen (m pl)
carros (m pl) y carretillas (f pl)
chariots (m pl) et fardiers (m pl)
autocarri (m pl) e carrelli (m pl)

18 tubes
Röhren (f pl)
tubos (m pl)
tubes (m pl)
tubi (m pl)

1 tubes – layflat
Schläuche (m pl) – flach
tubos (m pl) layflat
tubes (m pl) de poser à plat
tubi (m pl) – orizzontali/tubi orizzontali

2 tubes – plastic
Schläuche (m pl) – Kunststoff (m)
tubos (m pl) de plástico
tubes (m pl) plastiques
tubi (m pl) di plastica/tubi di plastica

3 tuft
Troddel (f)/Quaste (f)
Tuft
tuft (m)
ciuffetto (m)

4 turnover
Umsatz (m)
volumen de ventas
chiffre (m) d'affaires
giro (m) d'affari

5 turns per inch
Drehungen (f pl) pro Zoll
vueltas (f pl) por pulgada
tours (m pl) par pouce
giri (m pl) per pollice

6 twill leave
Köperbinbing (f), Twill (m), Serge (m)
tejer (m) asargado
armure (f) sergée
armatura (f) diagonale

7 twines
Bindfaden (m pl)
guitas (f pl)
ficelles (f pl)
spaghi (m pl)

8 twist
Drehung (f)
torsión (f)
torsion (f)
torsione

9 two ply
Zweifach-, Doppel-,
de dos capas
à deux brins
ritorto (m) di due capi

10 tying-in
Anknüpfen (n)
ligar (m) /unir
nouage (m)
legare

U

11 ultrasonic
Ultraschall (m)
ultrasonidos (m pl)
ultrason (m)
scienza (f) degli ultrasuoni/ultrasuoni

12 undyed
ungefärbt
no teñido
non teint
non tinto

13 uniform [standard]
einheitlich
uniforme
uniforme
uniforme [standard]

14 unrolling
Abwickeln (n)
desenrollar (m)
travail (m) à la déroulée
srotolamento (m)

15 upholstery
Polsterware (f)
tapiceria
tapisserie (f) d'ameublement
tappezzeria (f)

16 used [second-hand]
gebraucht, Gebraucht-
de ocasión
d'occasion
usato [di seconda mano]

17 vacuum
Vakuum (n)
vacio (m)
vide (m)
vuoto (m)

1 variable speed
variable Geschwindigkeit (f)
velocidad (f) variable
vitesse (f) réglable constante
velocità (f) variable

2 vat
Küpe (f)
tina (f)
cuve (f)
tino (f)

3 vat dyeing
Küpenfärbung (f)
teñir (m) en tina
teinture (f) aux colorants de cuve
colorazione (f) a tino/tintura (f) al tino (m)

4 V-belt
V-Riemen (m)
correa (f) de transmisión
courroie (f) trapézoîdale
cinghia (f) trapezoidale

5 vegetable oil
Pflanzenöl (n)
aceite (m) vegetal
huile (f) végétale
olio (m) vegetale

6 ventilation systems
Ventilationssysteme (n pl)
sistemas (m pl) de ventilación
installations (f pl) d'aérage
sistemi (m pl) di ventilazione

7 vessel
Gefäß (n)
recipiente
récipient (m)
recipiente (m)

8 video service
Videoservice
servicios (m pl) de vídeo
services (m pl) vidéo
servizi video

9 virgin wool
Schurwolle (f)
lana (f) virgen
lain (f) vierge
lana (f) vergine

10 viscose
Viskose (f)
viscosa (f)
viscose (f)
viscosa (f)

11 volume
Volumen (n)
volumen (m)
volume (m)
volume (m)

12 wadding
Wattierung (f)
entretela (f)
rembourrage (m)
ovatta (f)

13 waddings
Wattierungen (f pl)
entretelas (f pl)
rembourrages (m pl)
ovatta (f pl) per imbottitura

14 warehouse
Warenlager (n)
almacén (m)
entrepôt (m)
magazzino (m)

15 warehousing/storage
Lagern (n)
almacenaje (m)
mise (f) en dépôt
immagazzinamento (m)

16 warp
Webkette (f)
urdimbre (m)
chaîne (f) de tissage
ordito (m)

17 warp dressers
Kettschlichter (m pl)
aderezadores (m pl) de urdimbre
machines (f pl) à dresser les chaînes
apprettatrici (f pl) dell'ordito/ apprettatori dell'ordito

18 warp - dressings
Ketten (f pl) – Schlichtenmittel (f pl)
aderezos (m pl) de urdimbre
montage (m) de la chaîne
ordito (m) – imbozzimatura

1 warp – knitters
Ketten (f pl) – Wirker (m pl)
urdimbre (f) para hacer punto
métiers (m pl) chaîne
ordito (m) – maglieri (m pl)

2 warp lubricants
Kettengleitmittel (n pl)
lubricantes (m pl) para urdimbre
lubrifiants (m pl) de chaîne
lubrificanti (m pl) dell'ordito

3 warp waxes
Kettenwachs (n)
ceras (f pl) para urdimbre
cires (f pl) pour chaîne
cere (f pl) per ordito

4 warper
Schärmaschine (f)
urdidor (m)
ourdissoir (m)
orditore (m)

5 warping
Schären (n)
urdir (m)
ourdissage (m)
orditura (f)

6 warp knitting
Kettenwirkerei (f)
tejido de (m) punto por urdimbre
tricotage (m) chaîne
tessitura (f) a maglia in catena

7 warp knotting
Kettenknüpferei (f)
nudo (m) de urdimbre
nouage (m) de la chaîne
annodare (v) l'ordito (annoclare dell'ordito)

8 waste
Abfall (m)
residuos (m pl)
déchets (m pl)
scarto/cascame

9 waste – alpaca
Ausschuß – Alpaka (n)
residuos (m pl) de alpaca
déchets (m pl) – alpaga
scarto (m) – alpaca

10 waste – carpet yarn
Ausschuß (m) – Teppichgarn (m)
residuos (m pl) de hilo de alfombras
déchets (m pl) – fil de tapis
scarto (m) – filato per tappeti

11 waste – combing
Abfallkämmen (n)
peinado (m) de residuos
entredents (m pl)/déchet de peignage
scarto (m) – pettinatura

12 waste – cotton
Abfall (m) - Abfallbaumwolle (f)
residuos (m pl) de algodón
déchets (m pl) de coton
scarto (m) – cotone

13 waste – cotton, cleaning
Abfall (m) - Putzwolle (f)
residuos (m pl) limpieza de algodón
coton (m) à polir
scarto (m) – pulitura del cotone

14 waste – cutters
Ausschuß (m) – Schneider (m pl)
cortadoras (f pl) de residuos
déchets (m pl) – découpeuses
scarto (m) – tagliatrici

15 waste – exporters
Abfall (m) - Abfallexporteurs (m pl)
exportadores (m pl) de residuos
exportateurs (m pl) de déchets
scarto (m) – esportatori

16 waste – for synthetic regeneration
Auschuß (m) – für Synthetik-regenerierung Regenerat
residuos (m pl) para recuperación de sintéticos
déchets (m pl) – pour récupération synthétique
scarto (m) – per rigenerazione sintetica

17 waste – garnetters
Abfall (m) - Abfallrückgewinner (m pl)
arrancadores (m pl) de residuos
garnetteurs (m pl) de déchets
scarto (m) di garnettatrici

18 waste – hosiery
Abfall (m) - Strick- und Strumpfwarenabfall (m)
residuos (m pl) de calcetería
déchets (m pl) de bonneterie
scarto (m) di calzetteria (f)

1 waste – importers
Abfall (m) - Abfallimporteure (m pl)
importadores (m pl) de residuos
importateurs (m pl) de déchets
scarto (m) d'importatori

2 waste – merchants
Abfall (m) - Abfallhändler (m pl)
comerciantes (m pl) de residuos
négociants (m pl) en déchets
scarto (m) dei commercianti

3 waste – mohair
Abfall (m) - Mohairabfall (m)
residuos (m pl) de angora
déchets (m pl) de mohair
scarto (m) – mohair

4 waste – nylon
Abfall (m) - Nylonabfall (m)
residuos (m pl) de nilón
déchets (m pl) de nylon
scarto (m) – nailon

5 waste – openers
Abfall (m) - Abfallöffner (m pl)
abridores (m pl) de residuos
effilocheuses (f pl) pour déchets et fils
scarto (m) – apritoi

6 waste – plastics
Abfall (m) - Plastikabfall (m)
residuos (m pl) de plásticos
déchets (m pl) plastiques
scarto (m) di plastica

7 waste – polyester
Auschuß (m) – Polyester (n)
residuos (m pl) de poliéster
déchets (m pl) – polyester
scarto (m) di poliestere

8 waste – polypropylene
Abfall (m) - Polypropylenabfall (m)
residuos (m pl) de polipropileno (m)
déchets (m pl) de polypropylène
scarto (m) di polipropilene

9 waste – processors
Abfall (m) - Abfallverarbeiter (m pl)
residuos (m pl) de elaboración
machines (f pl) à traiter les déchets
scarto (m) di trasformatrici

10 waste – pullers
Abfall (m) - Abfallzieher (m pl)
residuos (m pl) para separar los hilos
machines (f pl) à séparer les fils durs des déchets
scarto (m) – tiratoi

11 waste – pullers, all–wool
Ausschuß (m) – Verzieher, alles Wolle
tiradores (m pl) de residuos de lana pura
déchets (m pl) – délaineurs purls laines
scarto (m) – tiratoi per lana pura

12 waste – pullers hosiery clips
Ausschuß (m) – Strumpfklammern (f pl)
pinzas de calcetería para tiradores de residuos
déchets (m pl) – pinces arracheuses pour bonneterie
scarto (m) – tiratoi – pinzette da maglieria

13 waste – pullers laps and spinning
Ausschuß (m) – Verzieher, Wickel und Spinnen
residuos (m pl) de hilar y de rodillos
déchets (m pl) – arracheuses – nappes et filature
scarto (m) – tiratoi – tela di battitoio e filatura

14 waste – pullers wool/synthetic
Ausschuß (m) – Verzieher, Wolle/Synthetik
tiradores (m pl) de residuos de lana/sintéticos
déchets (m pl) – arracheuses – laine/synthétiques
scarto (m) – tiratoi – lana/sintetici

15 waste – silk
Abfall (m) - Seidenabfall (m)
residuos (m pl) de seda
déchets (m pl) de soie
scarto (m) di seta

16 waste – sorters
Abfall (m) - Abfallsortierer (m pl)
clasificadores (m pl) de residuos
assortisseuses (f pl) de déchets
scarto (m) di selezionatrici

1 waste – spinning and hosiery
Abfall (m) - Spinn und Wirkwaren-
abfall (m)
residuos (m pl) de hilar y de calcetería
déchets (m pl) de filature et bonneterie
scarto (m) di filatura e maglieria

2 waste – synthetics
Abfall (m) - Synthetikabfall (m)
residuos (m pl) de sintéticos
déchets (m pl) synthétiques
scarto (m) di materie sintetiche

3 waste – willeyers
Abfall (m) - Abfallreißwölfe (m pl)
residuos (m pl) de paño denso
batteurs (m pl) de déchets
scarto (m) – diavolotti

4 waste – wipers
Abfall (m) – Abfallwischtücher (n pl)
residuos (m pl) para limpieza
machines (f pl) – torchon de déchets
scarto (m) – strofinacci

5 waste – wool
Abfall (m) – Wollabfall (m)
residuos (m pl) de lana
bourre (f) de laine
scarto (m) di lana

6 waste – woollen
Abfall – wollener Abfall (m)
residuos (m pl) de lana
déchets (m pl) de laine
scarto (m) – di lana

7 waste – worsted
Abfall (m) – Kammgarnabfall (m)
residuos (m pl) de estambre
déchets (m pl) de laine peignée
scarto (m) di pettinato

8 waste – yarn
Abfall (m) – Garnabfall (m)
residuos (m pl) de hilo
déchets (m pl) de filé
scarto (m) di filato

9 water
Wasser (n)
agua (f)
eau (f)
acqua (f)

10 water jet loom
Wasserstrahl-Düsenwebstuhl (m)
telar (m) de toberas de agua
métier à jet d'eau
macchina (f) da tessere ad ugello

11 waterproofing
wasserdichte Appretierung (f)
impermeabilización
imperméabilisation (f)
impermeabilizzazione (f)

12 water treatment
Wasserbehandlung (f)
tratamiento (m) de agua
traitement (m) des eaux
trattamento (m) dell'acqua

13 wax
Wachs (n)
cera (f)
cire (f)
cera (f)

14 wax lubricants
Wachsgleitmittel (n pl)
lubricantes (m pl) de cera
lubrifiants (m pl) à la cire
lubrificanti (m pl) a cera/di cera

15 wax rings
Wachsringe (m pl)
anillos (m pl) de cera
anneaux (m pl) de paraffine
anelli (m pl) di cera

16 weak
schwach, verdünnt
débil/flojo
faible
debole

17 weave
Webart (f)
tejido (m)
tisser (v)
armatura (f)

18 weaver
Weber (m)/Weberin (f)
tejedor (m)
tisseur (m)/tisseuse (f)
tessitore (m) tessitrice (f)

1 weaving
Weben (n)
tejer (m)
tissage (m)
tessitura (f)/tessitore (m)

2 weft
Füllung (f)/Schuß (m)
trama (f)
trame (f)
trama (f)

3 weight
Gewicht (n)
peso (m)
poids (m)
peso (m)

4 wet
naß
húmedo
mouillé
bagnato

5 wet finishing
Naßappretur (f)
acabado (m) con agua
apprêt (m) au mouillé
finissaggio (m) bagnato

6 white
weiß
blanco
blanc
bianco

7 width
Weite (f), Breite (f)
anchura (f)
largeur (f)
larghezza (f)

8 willeyers, fearnoughters and blenders
Reiß- und Krempelwölfe (m pl) und Mischmaschinen (f pl)
fabricantes (m pl) de paño denso y mezcladores
batteurs (m pl), effilocheuses (f pl) et mélangeuses (f pl)
diavolotti (m pl), sfibratrici (f pl) e mescolatrici (f pl)

9 wind (v)
wickeln
devanar
enrouler, dévider
incannare

10 winding
Wickeln (n)
desvanar (m)
bobinage (m)
incannatura (f)

11 wool
Wolle (f)
lana (f)
laine (f)
lana (f)

12 wool agents
Wollmittel (n pl)
agentes (m pl) de lana
agents (m pl) en laine
agenti (m pl) di lana

13 wool – all origins
Wolle (f) – verschiedenen Ursprungs
lana (f) de varias procedencias
laine (f) de toutes provenances
lana (f) – di origine varia/di provenienza varia

14 wool – auctioneers
Wolle (f) – Wollversteigerer (m pl)
subastadores (m pl) de lana
directeurs (m pl) de la vente aux enchères de la laine
banditori di lana

15 wool – blends
Wolle (f) – Wollmelange (f)
mezcla (f) de lanas
mélange (m) de laine
mischie di lana

16 wool – blending
Wolle (f) – Wollmischen (n)
mezcla (m) de lanas
mélange (m) de laine
mischie di lana

17 wool – British
Wolle (f) – britische Wolle (f)
lana (f) británica
laine (f) britannique
lana (f) – inglese

18 wool – brokers
Wolle (f) – Wollmakler (m pl)
corredores (m pl) de lana
courtiers (m pl) en laine
mediatori di lana

1 wool – carboised, carbonized
Wolle (f) – karbonisierte Wolle (f)
lana (f) carbonizada
laine (f) carbonisée
lana (f) – carbonizzata

2 wool – carpet
Wolle (f) – Wollteppich (m), Teppichwolle (f)
lana (f) para alfombras
laine (f) pour tapis
tappeti di lana

3 wool – classing
Wolle (f) – Wollklassierung (f)
clasificación (f) de lana
classement (m) de laine
classificazione di lana

4 wool – combers
Wolle (f) – Wollkämmer (m pl)
peinadores (m pl) de lana
peigneurs (m pl) de laine
pettinatrici di lana

5 wool – decotters
Wolle (f) – Wollentfilzer (m pl)
extractores (m pl) de chavetes de lana
débourreurs (m pl) de laine
lana (f) – estrattori di ciuffi di lana

6 wool – deskinners
Wolle (f) – Wollenthäuter (m pl)
despellejadores (m pl) de lana
délaineurs (m pl) de laine
lana (f) – depilatore per lana

7 wool – English
Wolle (f) – englische Wolle (f)
lana (f) inglesa
laine (f) anglaise
lana (f) – inglese

8 wool – exporters
Wolle (f) – Wollexporteure (m pl)
exportadores (m pl) de lana
exportateurs (m pl) de la laine
esportatori di lana

9 wool – grading
Wolle (f) – Wollklassierung (f)
clasificación (f) de lana
classification (f) de la laine
classificazione di lana

10 wool – importers
Wolle (f) – Wollimporteure (m pl)
importadores (m pl) de lana
importateurs (m pl) de la laine
importatori di lana

11 wool – karakul
Wolle (f) – Krimmer (m)
lana (f) de astracán
laine (f) – imitation d'astrakan
lana (f) – karakul

12 wool – karakul, bleached
Wolle (f) – Krimmer, gebleicht
lana (f) blanqueada de astracán
laine (f) – imitation d'astrakan blanchie
lana (f) – karakul – candeggiata

13 wool – karakul, scoured
Wolle (f) – Krimmer, gewaschen
lana (f) desengrasada de astracán
laine (f) – imitation d'astrakan lavée à fond
lana (f) – karakul – lavata

14 wool – kemps
Wolle (f) – Grannenhaar
fibras (f pl) huecas de lana
laine (f) jarreuse
lana (f) – fibra robusta

15 wool – lambs'
Wolle (f) – Lammwolle (f)
lana (f) de cordero
laine (f) d'agneau
lana (f) d'agnello

16 wool – lambs' carbonised, carbonized
Wolle (f) – Lamm, karbonisiert
lana (f) carbonizada de cordero
laine (f) d'agneau carbonisée
lana (f) d'agnello – carbonizzata

17 wool – lambs', scoured
Wolle (f) – Lamm, gewaschen
lana (f) desengrasada de cordero
laine (f) d'agneau lavée à fond
lana (f) – agnello – lavata

18 wool – merchants'
Wolle (f) – Wollhändler (m pl)
comerciantes (m pl) de lana
négociants (m pl) en laine
commercianti di lana

1 wool – openers
Wolle (f) – Wollöffner (m pl)
abridores (m pl) de lana
machines (f pl) d'ouverture de la laine
apritoi di lana

2 wool – reclaimers
Wolle (f) – Reißwolle
renovadores (m pl) de lana
rénovateurs (m pl) de laine
recuperatori di lana

3 wool – scoured and carbonised, carbonized
Wolle (f) – gewaschen und karbonisiert
lana (f) desengrasada y carbonizada
laine (f) lavée à fond et carbonisée
lana (f) – lavata e carbonizzata

4 wool – scoured, New Zealand
Wolle (f) – gewaschen, Neuseeland
lana (f) desengrasada de Nueva Zelanda
laine (f) de Nouvelle-Zélande lavée à fond
lana (f) – lavata – Nuova Zelanda

5 wool – scourers
Wolle (f) – Wollwäscher (m pl)
desengrasadores (m pl) de lana
machines (f pl) de lavage à chaud de la laine
lavatrici di lana

6 wool – Shetland
Wolle (f) – Shetlandwolle (f)
lana (f) de Shetland
laine (f) Shetland
lana (f) – shetland

7 wool – slipe
Wolle (f) – Gerberwolle (f)
lana (f) de piel tratada con cal
laine (f) de peau traitée à la chaux
lana (f) – calcinata

8 wool – sorters
Wolle (f) – Wollsortierer (m pl)
clasificadores (m pl) de lana
trieuses (f pl) de laine
selezionatrici di lana

9 wool – sorting
Wolle (f) – Wollsortieren (n)
clasificación (f) de lana
triage (m) de laine
lana (f) – cernita

10 wool – spinning
Wolle (f) – Wollspinnen (n)
hilatura (f) de lana
filaturE (f) de laine
lana (f) – filatura

11 wool – willyers and blenders
Wolle (f) – Reißer und Mischer
limpiadores (m pl) y mezcladores (m pl) de lana
mélangeuses (f pl) et ouvreuses (f pl) de laine
lana (f) – diavolotti e mescolatrici

12 wool – yak
Wolle (f) – Yakwolle (f)
lana (f) de yak
laine (f) de yak
lana (f) – yak

13 wool – yak, dehaired
Wolle (f) – Yak, enthaart
lana (f) pelada de yak
laine (f) de yak débourrée
lana (f) – yak – privo di pelo

14 Woollen and Worsted cloth merchants
Woll- und Kammgarntuchhändler (m pl)
comerciantes (m pl) de telas de lana y de estambre
négociants (m pl) en tissu de laine et tissu de laine peignée
commercianti (m pl) di tessuti di lana e pettinati

15 Woollen and Worsted yarn merchants
Woll und Kammgarnhändler (m pl)
comerciantes (m pl) de hilos de lana y de estambre
négociants (m pl) en fil de laine cardée et peignée
commercianti (m pl) di tessuti di lana e filati pettinati

16 Woollen cloth manufacturers
Wolltuchhersteller (m pl)
fabricantes (m pl) de tela de lana
fabricants (m pl) de tissu de laine
produttori (m pl) di tessuti di lana

1 Woolen cloth manufacturers – blended/woolmark
Wolltuchhersteller (m pl) – Mischwolle/Wollsiegel
fabricantes (m pl) de tela de lana mezclada
fabricants (m pl) de tissu de laine mélangée
produttori (m pl) di tessuti di lana – misti

2 Woollen cloth manufacturers – speciality fibres
Wolltuchhersteller (m pl) – Spezialfasern
fabricantes (m pl) de tela de lana de fibras especiales
fabricants (m pl) de tissu de laine – fibres spéciales
produttori (m pl) di tessuti di lana – fibre speciali

3 Woollen cloth manufacturers – woolmark
Wolltuchhersteller (m pl) – Wollsiegel
fabricantes (m pl) de tela de lana pura
fabricants (m pl) de tissu de laine – woolmark
produttori (m pl) di tessuti di lana – marchi di pura lana

4 Woollen spinning
Streichgarnspinnen (n)
hilatura (f) de lana
filature (f) de laine cardée
filatura (f) di tessuti di lana

5 Woollen yarn spinners
Wollgarnspinner (m pl)
hiladores (m pl) de hilo de lana
filateurs (m pl) de laine cardée
filatori (m pl) per filati di lana pettinata

6 Woollen yarn spinners – speciality fibres
Wollgarnspinner (m pl) – Spezialfasern
hiladores (m pl) de lana de fibras especiales
filateurs (m pl) de fil de laine cardée – fibres spéciales
filatoi (m pl) per filati di lana – fibre speciali

7 work study
Arbeitsstudie (f)
estudio (m) de organización del trabajo
étude (f) de travail
studio (m) del lavoro

8 Worsted and Mohair yarn spinners
Kammgarn und Mohairgarnspinner (m pl)
hiladores (m pl) de hilos de estambre y angora
filateurs (m pl) du peigné et du mohair
filatoi (m pl) per pettinati e filati di mohair

9 Worsted cloth manufacturers
Kammgarnstoffhersteller (m pl)
fabricantes (m pl) de tela de estambre
fabricants (m pl) de tissu de laine peignée
produttori (m pl) di tessuti pettinati

10 Worsted cloth manufacturers – fine and fancy
Kammgarngewebehersteller (m pl) – fein und mit Effekten
fabricantes (m pl) de tela de estambre – fina y adornada
fabricants (m pl) d'étoffe de laine peignée – fine et fantaisie
produttori (m pl) di tessuti pettinati – di prima qualità e fantasia

11 Worsted cloth manufacturers – super 80's plus
Kammgarngeweberhersteller (m pl) – Super 80 plus
fabricantes (m pl) de tela de estambre – super 80's plus
fabricants (m pl) d'étoffe de laine peignée – Super 80 plus
produttori (m pl) di tessuti pettinati – super 80 plus

12 Worsted cloth manufacturers – speciality fibres
Kammgarngewebehersteller (m pl) – Spezialfasern
fabricantes (m pl) de tela de estambre de fibras especiales
fabricants (m pl) d'étoffe de laine peignée – fibres spéciales
produttori (m pl) di tessuti pettinati – fibre speciali

1 Worsted yarn spinners – super 80's plus
Wollkammgarnspinner (m pl) – Super 80 plus
hiladores (m pl) de estambre – super 80's plus
filateurs (m pl) du peigné – Super 80 plus
filatori (m pl) di filati pettinati – super 80 plus

2 Worsted yarn spinners – speciality fibres
Wollkammgarnspinner (m pl) – Spezialfasern (f pl)
hiladores (m pl) de estambre – fibras especiales
filateurs (m pl) du peigné – fibres spéciales
filatori (m pl) di filati pettinati – fibre speciali

3 Worsted spinning
Kammgarnspinnen (n)
hilatura (f) de estambre
filature (f) du peigné
filatura (f) pettinata

4 Worsted yarn spinners
Kammgarnspinner (m pl)
hiladores (m pl) de hilo de estambre
filateurs (m pl) du peigné
filatori di filatura pettinata

5 Worsted yarn spinners – semi
Halbkammgarnspinner (m pl)
hiladores (m pl) de hilo de estambre con lana
filateurs (m pl) du mi-peigné
filatori di filatura semi-pettinata

6 woven
gewebt
tejido
tissé
tessuto

7 pleat/crease//fold
Falte (f), Knitterfalte (f)
arruga (f)
froisser (v)
piega (f)

X

8 X-rays
Röntgenstrahlen (m pl)
rayos (m pl) X
rayons (m pl) X
raggi (m pl) X

9 yak
Yak (n)
yak (m)
yak (m)
yak (m)

10 yarn
Garn (n)
hilo (m)
fil (m)
filato (m)

11 yarn – acrylic
Garn (n) – Acrylgarn (n)
hilo (m) acrílico
fil (m) d'acrylique
filato (m) – acrilico

12 yarn – acrylic blends
Garn (n) – Acrylgarnmischungen (f pl)
hilo (m) de mezclas acrílicas
fil (m) d'acrylique mélangé
filato (m) – misto acrilico

13 yarn – acrylic bright high bulk
Garn (n) – helles Acrylhochbauschgarn (n)
hilo (m) acrílico de gran volumen, brillante
fil (m) d'acrylique brillant gonflé
filato (m) – acrilico lucido ad alta voluminosità

14 yarn – acrylic bright regular
Garn (n) – helles Acrylnormgarn (n)
hilo (m) acrílico y regular, brillante
fil (m) d'acrylique brillant régulier
filato (m) – acrilico lucente regolare

15 yarn – acrylic cotton
Garn (n) – Acrylbaumwollgarn (n)
hilo (m) acrílico con algodón
fil (m) de coton acrylique
filato (m) – acrilico cotone

1 yarn – acrylic high bulk
Garn (n) – Acrylhochbauschgarn (n)
hilo (m) acrílico de gran volumen
fil (m) d'acrylique gonflé
filato (m) – acrilico ad alta voluminosità

2 yarn – acrylic matt regular
Garn (n) – mattes Acrylnormgarn (n)
hilo (m) mate acrílico y regular
fil (m) d'acrylique mat régulier
filato (m) – acrilico opaco regolare

3 yarn – acrylic nylon
Garn (n) – Acrylnylongarn (n)
hilo (m) acrílico con nilón
fil (m) d'acrylique nylon
filato (m) – acrilico nailon

4 yarn – acrylic open-end
Garn (n) – Acrylendlosgarn (n)
hilo (m) acrílico de extremo abierto
fil (m) d'acrylique open end
filato (m) – acrilico – filatura (f)

5 yarn – acrylic slub
Garn (n) – Acrylflammeneffektgarn (m)
hilo (m) de mechones acrílicos
fil (m) d'acrylique boutonneux
acrilico ingrossamento del filato

6 yarn – acrylic/wool
Garn (n) – Acrylwollgarn (n)
hilo (m) acrílico con lana
fil (m) de laine acrylique
filato (m) – acrilico lana

7 yarn – agents
Garn (n) – Garnmittel (n pl)
agentes (m pl) de hilos
agents (m pl) de fil
agenti di filato

8 yarn – alpaca
Garn (n) – Alpakagarn (n)
hilo (m) de alpaca
fil (m) d'alpaga
filato (m) – alpaca

9 yarn – alpaca/wool
Garn (n) – Alpaka/Wolle
hilo (m) de alpaca/lana
fil (m) d'alpaga/laine
filato (m) – alpaca/lana

10 yarn – angola
Garn (n) – Angolagarn (n)
hilo (m) de angola
fil (m) d'angola
filato (m) – angola

11 yarn – angora
Garn (n) – Angoragarn (n)
hilo (m) de angora
fil (m) d'angora
filato (m) – angora

12 yarn – Aran [knitting]
Garn (n) – Aranwirkgarn (n)
hilo (m) de hacer punto de aran
fil (m) Aran à tricoter
filato (m) – aran [lavorazione]

13 yarn – asbestos
Garn (n) – Asbest (n)
hilo (m) de amianto
fil (m) d'amiante
filato (m) – amianto

14 yarn – ballers
Garn (n) – Garnaufwickler (m pl)
ovilladores (m pl) de hilo
pelotonneuses (f pl)
imballatrici di filato

15 yarn – Berber
Garn (n) – Berbergarn (n)
hilo (m) berberisco
fil (m) Berber
filato (m) – berbero

16 yarn – blanket
Garn (n) – Deckengarn (n)
hilo (m) para mantas
fil (m) pour couvertures de lit
filato (m) – per coperte

17 yarn – blended
Garn (n) – Mischgarn (n)
hilo (m) mezclado
fil (m) mélangé
filato (m) – misto

18 yarn – bouclé, crepe, etc.
Garn (n) – Boucle-, Crepegarn usw
hilo (m) de bucle, crespón, etc.
bouclé (m), crêpe (m), etc.
filato (m) – boucle, crespo

1 yarn – British wool
Garn (n) – britisches Wollgarn (n)
hilo (m) de lana británica
fil (m) de laine britannique
filato (m) – lana inglese

2 yarn – brushed
Garn (n) – gebürstet
hilo (m) cepillado
fil (m) gratté
filato (m) – pettinato

3 yarn – camel hair
Garn (n) – Kamelhaargarn (n)
hilo (m) de pelo de camello
fil (m) de poil de chameau
filato (m) – pelo di cammello

4 yarn – camelhair/wool
Garn (n) – Kamelhaar/Wolle
hilo (m) de pelo de camello/lana
fil (m) de laine/poil de chameau
filato (m) – pelo di cammello/lana

5 yarn – carpet
Garn (n) – Teppichgarn (n)
hilo (m) para alfombras
fil (m) pour tapis
filato (m) – per tappeti

6 yarn – carpet woolmark
Garn (n) – Teppichwollsiegel (n)
hilo (m) de lana pura para alfombras
fil (m) de tapis woolmark
filato (m) – per tappeti – marchio di pura lana

7 yarn – carpet wool mixtures
Garn (n) – Teppichwollmischungen (f pl)
hilo (m) de mezclas de lana para alfombras
fil (m) de tapis – laine mélangée
filato (m) – misto lana per tappeti

8 yarn – cashmere
Garn (n) – Kaschmirgarn (n)
hilo (m) de cachemilon
fil (m) de cachemire
filato (m) – cashmere

9 yarn – cashmere/wool
Garn (n) – Kaschmir/Wolle
hilo (m) de cachemir/lana
fil (m) de laine/cachemire
filato (m) – cashmere/lana

10 yarn – cashmilon
Garn (n) – Kaschmilon (n)
hilo (m) de cachemir
fil (m) cashmilon
filato (m) – cashmere/nylon – cashmilon

11 yarn – chainette
Garn (n) – Chainettegarn (n)
hilo (m) de cadeneta
fil (m) chaînette
filato (m) – ritorto fantasia

12 yarn – chenille
Garn (n) – Chenillegarn (n)
hilo (m) de felpilla
fil (m) chenille
filato (m) – ciniglia

13 yarn – Cheviot [weaving]
Garn (n) – Cheviotwollgarn (n)
hilo (m) de tejer de cheviot
fil (m) cheviote
filato (m) – cheviot [tessitura]

14 yarn – conditioning agents
Garn (n) – Konditioniermittel (n)
agentes (m pl) para acondicionar hilo
agents (m pl) de conditionnement pour fils
agenti di condizionamento per filato

15 yarn – core-spun
Garn (n) – Seelengarn (n), Umspinnungsgarn (n)
hilo (m) hilado en torno de hilo nuclear
filé à fil d'âme
filato (m) – bobina/anima filata

16 yarn – cotton
Garn (n) – Baumwollgarn (n)
hilo (m) de algodón
fil (m) de coton
filato (m) di cotone

17 yarn – cotton/acrylic
Garn (n) – (Baumwollacrylgarn (n)) Baumwoll/Acrylgarn (n)
hilo (m) acrílico con algodón
fil (m) de coton/acrylique
filato (m) – cotone/acrilico

18 yarn – cotton blends
Garn (n) – Baumwollgarmischungen (f pl)
hilo (m) de mezclas de algodón
fil (m) de coton mélangé
filato (m) – misto cotone

1 yarn – cotton, carded
Garn (n) – Baumwollkardengarn (n)
hilo (m) de algodón cardado
fil (m) de coton cardé
filato (m) – cotone cardato

2 yarn – cotton, combed
Garn (n) – gekämmtes Baumwollgarn (n)
hilo (m) de algodón peinado
fil (m) de coton peigné
filato (m) – cotone pettinato

3 yarn – cotton, condenser
Garn (n) – Baumwollvorgarn (n)
hilo (m) condensador de algodón
fil (m) de coton imitation
filato (m) – cotone per carda

4 yarn – cotton, dyed
Garn (n) – gefärbtes Baumwollgarn (n)
hilo (m) de algodón teñido
fil (m) de coton teint
filato (m) – cotone tinto

5 yarn – cotton nylon
Garn (n) – Baumwoll/Nylongarn (n)
hilo (m) de algodón con nilón
fil (m) de coton/nylon
filato (m) – cotone/nailon

6 yarn – cotton open-end
Garn (n) – Baumwollendlosgarn (n)
hilo (m) de algodón de extremo abierto
fil (m) de coton open end
filato (m) – cotone ad un'estremità aperta

7 yarn – cotton open, end/polyester
Garn (n) – Baumwollendlos/Polyestergarn (n)
hilo (m) de poliéster con algodón de extremo abierto
fil (m) de coton open end/polyester
filato (m) – cotone ad un'estremità aperta/poliestere

8 yarn – cotton – polyester
Garn (n) – Baumwoll/Polyestergarn (n)
hilo (m) de algodón y poliéster
fil (m) de coton/polyester
filato (m) – cotone/poliestere

9 yarn – cotton slub
Garn (n) – Baumwolleffektgarn (n)
hilo (m) de mechones de algodón
fil (m) de coton boutonneux
ingrossamento di filato – cotone (m)

10 yarn – cotton speciality
Garn (n) – Baumwollspezialität (f)
hilo (m) especial de algodón
fil (m) de coton spécial
filato (m) – cotone speciale

11 yarn – cotton/viscose
Garn (n) – Baumwoll/Viskosegarn (n)
hilo (m) de algodón y viscosa
fil (m) de coton/viscose
filato (m) – cotone/viscosa

12 yarn – cotton/wool
Garn (n) – Baumwoll/Wollgarn (n)
hilo (m) de algodón y lana
fil (m) de coton/laine
filato (m) – cotone/lana

13 yarn – count
Garn (n) – Garnnummer (f)
tamaño (m) de hilo
tirre (m) de fil
filato (m) – titolo del filato

14 yarn – courtelle
Garn (n) – Courtellegarn (n)
hilo (m) courtelle
fil (m) courtelle
filato (m) – courtelle

15 yarn – craft
Garn (n) – Kunsthandwerksgarn (m)
hilo (m) de artesanía
fil (m) d'artisanat
filato (m) – d'arte

16 yarn – crochet
Garn (n) – Crochetgarn (n), Häkelgarn (n)
hilo (m) de ganchillo
fil (m) à crochet
filato (m) – per crochet

17 yarn – crossbred – hosiery
Garn (n) – Crossbredwirkgarn (n)
hilo (m) mestizo para calcetería
retors (m) pour bonneterie croisé
filato (m) – misto – per calzetteria

1 yarn – crossbred weaving
Garn (n) – Crossbredweben (n)
hilo (m) mulato para tejer
fil (m) à tisser croisé
filato (m) – incrociato per tessitura

2 yarn – curtain
Garn (n) – Vorhang/Stoffgarn (n)
hilo (m) para cortinas
fil (m) pour rideaux
filato (m) – per tende

3 yarn – Donegal
Garn (n) – Donelgarn (n)
hilo (m) donegal
fil (m) donegal
filato (m) – donegal

4 yarn – doublers
Garn (n) – Doubliergarn (n)
dobladores (m pl) de hilo
retordeuses (f pl) de fil
accoppiatrici del filato

5 yarn – dralon
Garn (n) – Dralongarn (n)
hilo (m) dralón
fil (m) dralon
filato (m) – dralon

6 yarn – dyed
Garn (n) – gefärbtes Garn (n)
hilo (m) teñido
teint en filés
filato (m) tinto

7 yarn – dyeing
Garn (n) – Garnfärben (n)
tintura (m) de hilo
teinture (f) en filés
tinto in filo

8 yarn – effect, fancy, etc.
Garn (n) – Effektgarn (n), Zierfaden (m pl) etc.
hilo (m) artístico, elegante, etc.
fil (m) d'effet, fantaisie, etc.
filato (m) – d'effetto, fantasia, ecc.

9 yarn – elastomeric
Garn (n) – Elastomergarn (n)
hilo (m) elástico
fil (m) élastomère
filato (m) – elastomerico

10 yarn – English wool
Garn (n) – englisches Wollgarn (n)
hilo (m) de lana inglesa
fil (m) de lana inglesa
fil (m) de laine anglaise
filato (m) – lana inglese

11 yarn – exporters
Garn (n) – Garnexporteure (m pl)
hilo (m) para la exportación
exportateurs (m pl) de fil
esportatori di filato/filato per l'esportazione

12 yarn – fabric – pile
Garn (n) – Gewebe (n) – Pol
hilo (m) para tela de pelusa
fil (m) pour tissu à poil
filato (m) – di lane incrociate – per tessere

13 yarn – filament
Garn (n) – Garn aus Endlosfäden (m pl), Endlosgern (n), Filamentgarn (n)
hilo (m) de filamentos
fil (m) continu
filato (m) – filamento

14 yarn – fine mink
Garn (n) – feiner Nerz (m)
hilo (m) fino de visón
fil (m) de vison fin
filato (m) – visone di qualità/filato di visone fine

15 yarn – flame-resistant
Garn (n) – feverfest
hilo (m) resistente al fuego
fil (m) ignifuge
filato (m) – resistente alla fiamma

16 yarn – flame-retardent
Garn (n) – flammernhemmend
hilo (m) ignífugo
fil (m) ignifuge
filato (m) – antifiamma

17 yarn – fur
Garn (n) – Pelz (m)
hilo (m) de piel
fil (m) pour fourrure
filato (m) – pelliccia

1 yarn – gaberdine
Garn (n) – Gabardinegarn (n)
hilo (m) para gabardinas
fil (m) gabardine
filato (m) – gabardine

2 yarn – gimps
Garn (n) – Gimpen (f pl)
hilo (m) de fantasía
fil (m) brodeur
filato (m) – fantasia

3 yarn – glass
Garn (n) – Glas (n)
hilo (m) de vidrio
fil (m) de verre textile
filato (m) – vetro

4 yarn – goat hair
Garn (n) – Ziegenhaargarn (n)
hilo (m) de pelo de cabra
fil (m) de poil de chèvre
filato (m) – pelo di capra

5 yarn – hair
Garn (n) – Haargarn (n)
hilo (m) de pelo
filé (m) de poil
filato (m) – pelo

6 yarn – handicraft – for macramé
Garn (n) – für Handarbeit (f) – für Macramé (n)
hilo (m) de artesanía para macramé
fil (m) d'artisanat pour macramé
filato (m) – lavorazione artigianale – per lavori a passamaneria (macramé)

7 yarn – hand-knitting
Garn (n) – als Handstrickgarn (n)
hilo (m) de hacer punto a mano
fil (m) à tricoter à la main
filato (m) – lavoro a maglia

8 yarn – hand-knitting – Aran
Garn (n) – als Aranhandstrickgarn (n)
hilo (m) aran de hacer punto a mano
fil (m) à tricoter à la main Aran
filato (m) – lavoro a maglia – aran

9 yarn – hand-knitting – ply
Garn (n) – als Handstrickzwin (m)
hilo (m) de capas de hacer punto a mano
retors (m) à tricoter à la main
filato (m) – lavoro a maglia – ritorto

10 yarn – heat-setting
Garn (n) – für Thermofixierung (f)
hilo (m) termofijación
thermofixage (m) du fil
filato (m) – di fissaggio a caldo

11 yarn – hosiery
Garn (n) – als Strumpf und Sockengarn (n)
hilo (m) para calcetería
retors (m) pour bonneterie
filato (m) – per calzetteria

12 yarn – importers
Garn (n) – für Garnimporteurs (m pl)
hilo (m) de importación
importateurs (m pl) de fil
importatori di filato

13 yarn – industrial
Garn (n) – als Industriegarn (n)
hilo (m) industrial
fil (m) industriel
filato (m) – industriale

14 yarn – interlining
Garn (n) – als Zwischenfuttergarn (n)
hilo (m) de entretelar
fil (m) pour entredoublure
filato (m) – per controfodere

15 yarn – Jersey
Garn (n) – als Jerseygarn (n)
hilo (m) de estambre fino
fil (m) de Jersey
filato (m) – jersey

16 yarn – knitted fabric – pile
Garn (n) – für Wirkgewebe (n) – Pol
hilo (m) para pelusa de tela de punto
fil (m) pour tissu à poil tricoté
filato (m) – tessuto lavorato a maglia – pelo

17 yarn – knitting
Garn (n) – als Strickgarn (n)
hilo (m) de hacer punto
fil (m) à tricoter
filato (m) – lavorazione a maglia

18 yarn – knitting-coloured
Garn (n) – als Farbstrickgarn (n)
hilo (m) de colores de hacer punto
fil (m) à tricoter coloré
filato (m) – lavorazione a maglia – colorato

1 yarn – knops
Garn (n) – als Noppengarn (n)
hilo (m) de borlitas
fil (m) avec boutons noueux
filato (m) – di bottoni

2 yarn – lambswool
Garn (n) – als Lammwollgarn (n)
hilo (m) de lana de cordero
fil (m) de laine d'agneau
filato (m) di lambswool/di lana d'agnello

3 yarn – lambswool/angora/nylon
Garn (n) – aus Lammwolle/Angora/Nylon
hilo (m) de lana de cordero/angora/nilón
fil (m) de laine d'agneau/angora/nylon
filato (m) – lambswool/angora/nailon

4 yarn – linen
Garn (n) – als Leinengarn (n)
hilo (m) de lienzo
fil (m) de lin
filato (m) di lino

5 yarn – linen/cotton
Garn (n) – als Leinenbaumwollgarn (n)
hilo (m) de lienzo/algodón
fil (m) de lin/coton
filato (m) di lino/cotone

6 yarn – linen – dry spun – line and tow
Garn (n) – für Leinen – Trockengesponnen – Langflachs und Towgarn
hilo (m) de lino hilado en seco – hilera y estopa
fil (m) de lin filé à sec – corde et étoupe
filato (m) di lino – filato a secco – lino e stoppa

7 yarn – linen – machine knitting
Garn (n) – als Leinen für Maschinenstrick
hilo (m) de lino para hacer punto a máquina
fil (m) de lin – tricot machine
filato (m) di lino – lavorazione a macchina

8 yarn – linen/viscose
Garn (n) – als Leinenviskosegarn (n)
hilo (m) de lienzo/viscosa
fil (m) de lin/viscose
filato (m) di lino/viscosa

9 yarn – linen – wet spun – line and tow
Garn (n) – für Leinen – naßgesponnen – Langflachs und Towgarn
hilo (m) de lino hilado con agua – hilera y estopa
fil (m) de lin filé au mouillé – corde et étoupe
filato (m) di lino – filato ad umido – lino e stoppa

10 yarn – linen/wool
Garn (n) – als Leinenwollgarn (n)
hilo (m) de lienzo/lana
fil (m) de lin/laine
filato (m) di lino/lana

11 yarn – loop
Garn (n) – als Schlingengarn (n)
hilo (m) de rizo
fil (m) bouclé
filato (m) – filato bouclé

12 yarn – lurex
Garn (n) – als Lurexgarn (n)
hilo (m) de lurex
fil (m) de lurex
filato (m) – lurex

13 yarn – manmade
Garn (n) – aus Kunstfaser (f)
hilo (m) sintético
fil (m) artificiel
filato (m) – artificiale/sintetico

14 yarn – marls
Garn (n) – gesprenkeltes Garn (n), Moulinégarn (n)
hilos (m pl) de colores mezclados
peigné (m) mélangé
filato (m) – marne/funi

15 yarn – melanges
Garn (n) – für Garnmischungen (f pl)
hilos (m pl) mezclados
fils (m pl) mélangés
filato (m) – melange

16 yarn – mercerised cotton
Garn (n) – merzerisiertes Baumwollgarn (n)
hilo (m) de algodón mercerizado
fil (m) de coton mercerisé
filato (m) – cotone mercerizzato

1 yarn – mercerised twist
Garn (n) – merzerisierte Zwirngarne (n pl)
hilo (m) de trenzas mercerizadas
retors (m) mercerisé
ritorto mercerizzato di filato

2 yarn – merchants
Garn (n) – Garnhändler (m pl)
comerciantes (m pl) de hilos
négociants (m pl) de fil
commercianti di filato

3 yarn – merino
Garn (n) – Merinowollgarn (n)
hilo (m) merino
fil (m) de mérinos
filato (m) – di pecore merino

4 yarn – merino/extra
Garn (n) – Merino/extrafein
hilo (m) de merino/extra
fil (m) de mérinos extra
filato (m) di lana merino/extra

5 yarn – mixture
Garn (n) – Mischgarn (n)
hilos (m pl) mezclados
fil (m) mixte
filato (m) di misto

6 yarn – mohair
Garn (n) – Mohairgarn (n)
hilo (m) de angora
fil (m) mohair
filato (m) di mohair

7 yarn – mohair/acrylic
Garn (n) – Mohair/Acrylgarn (n)
hilo (m) acrílico con angora
fil (m) mohair/acrylique
filato (m) di mohair/acrilico

8 yarn – mohair/wool
Garn (n) – Mohair/Wollgarn (n)
hilo (m) de angora/lana
fil (m) mohair/laine
filato (m) di mohair/lana

9 yarn – monofilament
Garn (n) – Monofilamentgarn (n)
hilo (m) de monofilamentos
fil (m) monofilament
filato (m) – monofilamento

10 yarn – mop
Garn (n) – Wischquastgarn (n)
hilo (m) para aljofifas
serpillière (f)
filato (m) – strofinaccio

11 yarn – natural
Garn (n) – Naturgarn (n)
hilo (m) natural
fil (m) naturel
filato (m) – naturale

12 yarn – nylon
Garn (n) – Nylongarn (n)
hilo (m) de nilón
fil (m) nylon
filato (m) di nailon

13 yarn – nylon sewing
Garn (n) – Nylonnähgarn (n)
hilo (m) de coser de nilón
fil (m) nylon à coudre
filato (m) di nailon – per cucire

14 yarn – nylon/wool
Garn (n) – Nylon/Wollgarn (n)
hilo (m) de nilón/lana
fil (m) de nylon/laine
filato (m) – nailon/lana

15 yarn – pile
Garn (n) – Pol
hilo (m) para pelusa
fil (m) à poil
filato (m) – pelo

16 yarn – polished
Garn (n) – Glanzgarn (n)
hilo (m) pulido
fil (m) glacé
filato (m) – lucidato

17 yarn – polyester
Garn (n) – Polyestergarn (n)
hilo (m) de poliéster
fil (m) polyester
filato (m) – poliestere

18 yarn – polyester/blends
Garn (n) – Polyestermischgarn (n)
hilo (m) de mezclas de poliéster
fil (m) polyester/mélanges
filato (m) – poliestere/misti

1 yarn – polyester/viscose
Garn (n) – Polyester/Viskosegarn (n)
hilo (m) de poliéster/viscosa
fil (m) polyester/viscose
filato (m) – poliestere/viscosa

2 yarn – polyester/wool
Garn (n) – Polyester/Wollgarn (n)
hilo (m) de poliéster/lana
fil (m) polyester/laine
filato (m) – poliestere/lana

3 yarn – polypropylene
Garn (n) – Polypropylengarn (n)
hilo (m) de polipropileno
fil (m) polypropylène
filato (m) – polipropilene

4 yarn – processors
Garn (n) – Garnverarbeiter (m pl)
elaboradores (m pl) de hilo
spécialistes (m pl) préposés au traitement du fil
filato (m) – trasformatrici

5 yarn – processors – autoclave heat setting and tumbling
Garn (n) – Verarbeiter – Autoklav-Thermofixierung und Tumblern
elaboradores (m pl) de hilo – ajuste de temperatura en autoclave y voltear
machines (f pl) de traitement (m) de fil – thermofixage par autoclave et à tambour
filato (m) – unità di lavorazione – posa a caldo e burattatura in autoclave

6 yarn – rare fibres/wool
Garn (n) – seltene Fasern/Wolle
hilo (m) de fibras exóticas/lana
fil (m) de laine/fibres rares
filato (m) – fibre preziose

7 yarn – redundant
Garn (n) – Überflüssiges Garn (n), Überschußgarn (n)
hilo (m) superfluo
fil (m) redondant
filato (m) – in eccesso/esuberante

8 yarn – repco
Garn (n) – Repco
hilo (m) repco
fil (m) repco
filato (m) – repco

9 yarn – ribbon
Garn (n) – Bandgarn (n)
hilo (m) para cintas
fil (m) ruban
filato (m) – nastro

10 yarn – rug
Garn (n) – Teppichgarn (n)
hilo (m) para alfombrillas
fil (m) pour tapis
filato (m) – per tappeto

11 yarn – rug – wool – ply
Garn (n) – Teppich (m) – Wolle (f) – Lage (f)
hilos (m pl) múltiples de lana para alfombrillas
fil (m) de laine simple pour tapis
filato (m) – tappeto – lana – ritorto (semplice)

12 yarn – rug – wool – 6 ply
Garn (n) – Teppich (m) – Wolle (f) – 6-lagig (f)
hilos (m pl) múltiples (6) de lana para alfombrillas
fil (m) de laine à six brins pour tapis
filato (m) – tappeto – lana – filato a 6 ritorti

13 yarn – saxony
Garn (n) – Elektoralgarn (n)
hilo (m) saxony
fil (m) Saxony
filato (m) – lana di sassonia

14 yarn – semi-worsted
Garn (n) – Halbkammgarn (n)
hilo (m) de estambre con lana
fil (m) mi-peigné
filato (m) – semi pettinato

15 yarn – semi-worsted – blended
Garn (n) – Halbkammgarn (n) – gemischt
hilo (m) mezclado – medio estambre
fil (m) de sayette mélangé
filato (m) – semi pettinato – misto

16 yarn – semi-worsted – coloured
Garn (n) – Halbkammgarn (n) – gefärbt
hilo (m) de colores – medio estambre
fil (m) de sayette teint
filato (m) – semi pettinato – colorato

1 yarn – semi-worsted – white
Garn (n) – Halbkammgarn (n) – weiß
hilo (m) blanco – medio estambre
fil (m) de sayette blanc
filato (m) – semi pettinato – bianco

2 yarn – sewing
Garn (n) – Nähgarn (n)
hilo (m) de coser
fil (m) à coudre
filato (m) – per cucine

3 yarn – sewing – polyester
Garn (n) – Nähen (n) – Polyester (n)
hilo (m) de poliéster para coser
fil (m) à coudre polyester
filato (m) – per cucire – poliestere

4 yarn – Shetland
Garn (n) – Shetlandgarn (n)
hilo (m) Shetland
fil (m) Shetland
filato (m) – shetland

5 yarn – Shetland – cotton/slub
Garn (n) – Shetland – Baumwolle/Vorgarn (n)
hilo (m) de Shetland – algodón/mechón
fil (m) de coton shetland noueux
filato (m) – shetland – cotone/bottone

6 yarn – Shetland – Donegal
Garn (n) – Shetland – Donegal
hilo (m) de Shetland y Donegal
fil (m) shetland Donegal
filato (m) – shetland – donegal

7 yarn – Shetland – featherweight
Garn (n) – Shetland – Federgewicht (n)
hilo (m) ligero de Shetland
fil (m) shetland léger
filato (m) – shetland – peso leggero

8 yarn – shoddy
Garn (n) – Shoddygarn (n)
hilo (m) de lana regenerada
fil (m) shoddy
filato (m) – lana rigenerata/lana shoddy

9 yarn – silk
Garn (n) – Seidengarn (n)
hilo (m) de seda
soie (f) filée
filato (m) – seta

10 yarn – silk bourette
Garn (n) – Seidenbourette
hilo (m) bourette de seda
soie (f) filée bourette
filato (m) – bovrette – seta/bourette di seta

11 yarn – silk/wool
Garn (n) – Seide/Wolle (f)
hilo (m) de seda/lana
fil (m) de laine/soie
filato (m) – seta/lana

12 yarn – sixfold
Garn (n) – Sechsfachgarn (n)
hilo (m) de seis capas
fil (m) sextuple
filato (m) – ritorto per maglieria a sei capi

13 yarn – slub
Garn (n) – Effektgarn (n), Flammengarn (n)
hilo (m) de mechones
fil (m) noueux
filato (m) – bottone

14 yarn – spinners
Garn (n) – Garnspinner (m pl)
hilo (m) hiladores
filateurs (m pl)
filato (m) – filatori

15 yarn – spun rayon
Garn (n) – gesponnenes Reyongarn (n)
hilo (m) de rayón hilado
rayonne (f) filée
filato (m) – raion filato

16 yarn – spun silk
Garn (n) – gesponnenes Seidegarn (n)
hilo (m) de seda hilada
fil (m) de déchets de soie
filato (m) – seta shappe (f)

17 yarn – stainless steel
Garn (n) – Edelstahl/Rostfreistahl (m)
hilo (m) de acero inoxidable
fil (m) acier inoxydable
filato (m) – acciaio inossidabile

18 yarn – stretch
Garn (n) – Stretchgarn (n)
hilo (m) elástico
fil (m) élastique
filato (m) – elasticizzato

1 yarn – strengthening agents
Garn (n) – Verstärkungsmittel (n pl)
agentes (m pl) para reforzar hilo
agents (m pl) de renforcement de fil
filato (m) – agenti rinforzanti

2 yarn – striping
Garn (n) – Ringelgarn (n)
hilo (m) de franjas
fil (m) de rayage
filato (m) – rigatura

3 yarn – super-carded
Garn (n) – superkardiert
hilo (m) super cardado
fil (m) cardé super
filato (m) – super cardato

4 yarn – supersoft knitting
Garn (n) – Superweichstricken (n)
hilo (m) super blando para hacer punto
fil (m) à tricoter très doux
filato (m) – estremamente soffice – per lavorare a maglia

5 yarn – synthetic
Garn (n) – Synthetikgarn (n)
hilo (m) sintético
fil (m) synthétique
filato (m) – sintetico

6 yarn – synthetic/wool
Garn (n) – Synthetik/Wollgarn (n)
hilo (m) dintético/lana
fil (m) de laine/synthétique
filato (m) – sintetico/lana

7 yarn – tweed
Garn (n) – Tweedgarn (n)
hilo (m) de dos colores
fil (m) tweed
filato (m) – tweed

8 yarn – twist
Garn (n) – Zwirngarn (n)
hilo (m) de torcer
retors (m)
filato (m) – ritorto

9 yarn – upholstery
Garn (n) – Polsterwarengarn (n)
hilo (m) para tapicería
fil (m) pour meubles
filato (m) – per tappezzeria

10 yarn – viscose
Garn (n) – Viskosegarn (n)
hilo (m) de viscosa
fil (m) de viscose
filato (m) – viscosa

11 yarn – wall-covering
Garn (n) – Wandbehanggarn (n), Wandtapetengarn (n)
hilo (m) para revestimientos de paredes
fil (m) de revêtement mural
filato (m) – per muri

12 yarn – waste
Garn (n) – Abfallgarn (n)/Garnabfall (m)
hilo (m) de residuos
déchets (m pl) de fil
filato (m) – scarto

13 yarn – water-soluble
Garn (n) – wasserlöslich
hilo (m) soluble en agua
fil (m) soluble dans l'eau
filato (m) – solubile in acqua

14 yarn – weaving
Garn (n) – Webgarn (n)
hilo (m) para tejer
fil (m) à tisser
filato (m) – per tessitura

15 yarn – weft knitting
Garn (n) – Kulierwirkgarn (n)
trama (f) para hacer punto
fil (m) à tricot cueilli
filato (m) – lavorazione a maglia in trama

16 yarn – winders
Garn (n) – Garnwickler (m pl)
devanadores (m pl) de hilo
dévidoirs (m pl) de fil
filato (m) – avvolgitrici

17 yarn – wool
Garn (n) – Wollgarn (n)
hilo (m) de lana
laine (f) filée
filato (m) – lana

18 yarn – wool/cotton
Garn (n) – Woll/Baumwollgarn (n)
hilo (m) de lana/algodón
fil (m) de laine/coton
filato (m) – lana/cotone

1 yarn – wool/nylon
Garn (n) – Woll/Nylongarn (n)
hilo (m) de lana/nilón
fil (m) de laine/nylon
filato (m) – lana/nailon

2 yarn – wool/nylon/acrylic
Garn (n) – Woll/Nylon/Acrylgarn (n)
hilo (m) de lana/nilón acrílico
fil (m) de laine/nylon/acrylique
filato (m) – lana/nailon/acrilico

3 yarn – wool/viscose
Garn (n) – Woll/Viskosegarn (n)
hilo (m) de lana/viscosa
fil (m) de laine/viscose
filato (m) – lana/viscosa

4 yarn – woollen
Garn (n) – wollens Garn (n)
hilo (m) de lana
fil (m) de laine cardée
filato (m) – di lana

5 yarn – woollen – blended
Garn (n) – wollenes Mischgarn (n)
hilo (m) de lana mezclada
fil (m) de laine cardée mélangé
filato (m) – di lana – misto

6 yarn – woollen – coloured
Garn (n) – gefärbtes Wollgarn (n)
hilo (m) de lana teñida
fil (m) de laine cardée coloré
filato (m) – di lana – colorato

7 yarn – woollen – hosiery
Garn (n) – wollenes Strumpfgarn (n)
hilo (m) de lana para calcetería
fil (m) de laine cardée pour bonneterie
filato (m) – di lana – calzetteria

8 yarn – woollen – knitting
Garn (n) – wollenes Strickgarn (n)
hilo (m) de lana para hacer punto
fil (m) de laine cardée à tricoter
filato (m) – di lana – per lavorare a maglia

9 yarn – woollen – natural
Garn (n) – Naturwollgarn (n)
hilo (m) de lana natural
fil (m) de laine cardée naturel
filato (m) – di lana – naturale

10 yarn – woollen – weaving
Garn (n) – wollenes Webgarn (n)
hilo (m) de lana para tejer
fil (m) de laine cardée à tisser
filato (m) – di lana – tessitura

11 yarn – woollen – white
Garn (n) – weißes Wollgarn (n)
hilo (m) de lana blanca
fil (m) de laine cardée blanc
filato (m) – di lana – bianco

12 yarn – woollen – woolmark
Garn (n) – Wollsiegelgarn (n)
hilo (m) de lana pura y clasificada
fil (m) de laine cardée woolmark
filato (m) – di lana – marchio di pura lana

13 yarn – worsted
Garn (n) – Kammgarn (m)
hilo (m) de estambre
fil (m) – de laine peignée
filato (m) – pettinato

14 yarn – worsted – blended
Garn (n) – gemischtes Kammgarn (n)
hilo (m) de estambre mezclado
fil (m) de laine peignée mélangé
filato (m) – pettinato – misto

15 yarn – worsted – coloured
Garn (n) – gefärbtes Kammgarn (n)
hilo (m) de estambre teñido
fil (m) de laine peignée coloré
filato (m) – pettinato – colorato

16 yarn – worsted – hosiery
Garn (n) – Kammgarnwirkwarengarn (n)
hilo (m) de estambre para calcetería
fil (m) de laine peignée pour bonneterie
filato (m) – pettinato – calzetteria

17 yarn – worsted – knitting
Garn (n) – Kammgarnstrickgarn (n)
hilo (m) de estambre para hacer punto
fil (m) de laine peignée à tricoter
filato (m) – pettinato per lavorare a maglia

18 yarn – worsted – semi-blended
Garn (n) – gemischtes Halbkammgarn (n)
hilo (m) de estambre mezclado con lana
fil (m) de laine peignée mi-mélangé
filato (m) – semi pettinato – misto

19 yarn – worsted – semi-coloured
Garn (n) – gefärbtes
hilo (m) de estambre y lana teñidos y mezclados
fil (m) de laine peignée mi-coloré
filato (m) – semi pettinato – colorato

20 yarn – worsted – semi-white
Garn (n) – weißes Halbkammgarn (n)
hilo (m) blanco de estambre mezclado con lana
fil (m) de laine peignée mi-blanc
filato (m) – semi pettinato – bianco

21 yarn – worsted – weaving
Garn (n) – Kammgarnwebgarn (n)
hilo (m) de estambre para tejer
fil (m) de laine peignée à tisser
filato (m) – pettinato – tessitura

22 yarn – worsted – white
Garn (n) – weißes Kammgarn (n)
hilo (m) blanco de estambre
fil (m) de laine peignée blanc
filato (m) – pettinato – bianco

23 yarn – wrap-spun (hollow spindle)
Garn (n) – Wickelspinngarn (n), Urnwindegarn (m)
hilo (m) hilado con aguja hueca
fil (m) guipe/
filato (m) – filati avvolti (fuso cavo)

24 yarn – yak
Garn (n) – Yakgarn (n)
hilo (m) de yak
fil (m) de yak
filato (m) – yak

25 yellow
gelb (adj)
amarillo
jaune
giallo

26 yield
Ertrag (m)
rendimiento (m)
rendement (m)
rendimento (m)

Z

27 Z twist
Z-Drehung (f)
torcedura (f) Z
torsion (f) Z
direzione (f)/torsione

Deutsch
German
Aléman
Allemand
Tedesco

arbeitsstudie (f) **152**, 7
arm (m) **9**, 15
arras-Wandteppich (m) **9**, 18
arsen (n) **10**, 1
artikel (m), Produkt (n), Warenstück (n) **10**, 2
asbest (m) **10**, 9
asbestfaser (f) **10**, 11
asbestgarne (n pl) **10**, 12
asbestgewebe (n pl) **10**, 10
astrachan **10**, 17
atmosphäre (f) **10**, 18
atmosphärendruck (m) **11**, 2
ätzen, angreifen; anheften **11**, 4
ätznatron (n), Natriumhydroxyd (n) **26**, 13
aufbereitung (f), Vorbehandlung (f), Bereitstellung (f), Präparation (f),
aufeinanderfolgend (adj) **52**, 5
aufladung (f), Ladung (f) **27**, 14
auflösen **65**, 12
aufnehmen, absorbieren **1**, 11
aufrauhen (n) **122**, 7
aufschlagen (n) **15**, 8
aufsichtsratsitzung (f) **19**, 5
aufstecken (n) **58**, 5
aufwickelvorrichtung (f) **139**, 11
auktion (f) **11**, 7
auktionator (m) **11**, 8
ausbessern (n) **120**, 14
ausbildung (f) **78**, 11
ausbrennen (n) **21**, 17
auschu (m) – Polyester (n) **147**, 7
auschu (m) – für Synthetikregenerierung Regenerat **146**, 16
ausführlich, eingehend, spezifiziert (adj) **64**, 3
ausgleich (m) Bilantz **13**, 2
ausgleichen, kompensieren **50**, 3
auspolstern, wattieren **126**, 4
ausrüsten, appretieren **67**, 14
ausschu – Alpaka (n) **146**, 9
ausschu (m) – Schneider (m pl) **146**, 14
ausschu (m) – Strumpfklammern (f pl) **147**, 12
ausschu (m) – Teppichgarn (m) **146**, 10
ausschu (m) – Verzieher, Wickel und Spinnen **147**, 13
ausschu (m) – Verzieher, Wolle/Synthetik **147**, 14
ausschu (m) – Verzieher, alles Wolle **147**, 11
aussehen (n), Bild (n) **8**, 14
ausstellen, aufweisen, vorführen, zeigen **65**, 11
auswahl (f) **134**, 5
auszählen **57**, 9
auszug (m) **80**, 3
autodachfütterungsstoff (m) Autodachfutterstoff (m) **36**, 5
autoklav (m) **11**, 10
automatisierung (f) **11**, 17
automatisch (adj) **11**, 13
automatischer Filmdruck (m) **11**, 14
automatischer Schützenwechsel (m) **11**, 15
automatisierte Steueranlagen (f pl) **11**, 12
axminsterteppich (m) **12**, 6
azetat (n) **2**, 7
azetatfarbstoff (m) **2**, 8
azetatfaser (f) **2**, 10
azetatgarn (n) **2**, 13
azetatgewebe (n) **2**, 9
azetatreyon (n), Azetatseide (f) **2**, 11
azetatzellwolle (f) **2**, 12
azeton (n) **2**, 15
azidität (f) **3**, 2
azinfarbstoff (m) **12**, 7
azofarbstoff (m) **12**, 8
azurblau (adj) **12**, 9

B

baby Blau (adj) **12**, 10
bad (n) **14**, 17
ballen (m) **13**, 4
ballenbänder (n pl) **13**, 6
ballenöffner (m) **13**, 5
band (n) **132**, 12
bänder (n pl) – Klebe-Bänder **139**, 14
bänder (n pl) – Polypropylen-Bänder **139**, 16
bänder (n pl) – Spindel-Bänder **139**, 17
bänder (n pl) – druckempfindliche Bänder **139**, 15
bänder (n pl) **139**, 13
bandware (f), Schmalgewebe (n) **122**, 5
bandwaren (f pl), Schmalgewebe (n pl) **122**, 6
bank (f) **13**, 12
bankfeiertag (m) **13**, 16
bankier (m) **14**, 3
bankkonto (n) **13**, 13
banknote (f) **13**, 18
bankrott (adj) **14**, 5
banküberweisungsuftrag (m) **14**, 4
barathea Gewebe (n) **14**, 6
baratheastoff (m) **31**, 18
barbary Schaf (n) **14**, 7
batschen (m) **14**, 15
batschenmaschine (f) **14**, 16
baumabdeckung (f) **15**, 2
bäumen (n) **15**, 5
baumfärber (m pl), Kettbaumfärber (m pl) **70**, 1
baumscheibe (f) **15**, 3
bäumvorrichtung (f) **15**, 6
baumwoll – Baumwollkardieren (n) **56**, 2
baumwoll- und Kunstfaser-Doublierung und -Seildrehen **55**, 16
baumwoll/Polyestergewebebleicher (m pl) **16**, 16

E

farbstoffe (m pl) für Transferdruck auf Wolle und Wollmischungen **78**, 7
farbstoffe (m pl) für Wolle **78**, 8
farbstoffe (m pl) für Zellulose **78**, 4
farbstoffe (m pl) und Pigmente (n pl) für Textildruck, Färbemittel und Pigments für Textildrucke **78**, 3
farbtonänderung (f), Farbenwechsel (m), Farbänderung (f) **44**, 18
faser (f) – Nylon (n), Nylonfaser (f) **82**, 7
faser (f) – Polyamid (m), Polymidfaser (f) **82**, 8
faser (f) – Polyester (m), Polyester faser (f) **82**, 9
faser (f) – Polypropylen (m), Polyproylenfaser (f) **82**, 10
faser (f) – Reyon, Rayfaser **82**, 11
faser (f) **81**, 17
faserschneider (m), Maschinen (f pl) **104**, 4
faservliesstoff (m) **38**, 9
fäulnisecht **133**, 6
feder (f) **137**, 16
feines Fell (n) **83**, 3
fell (n) **87**, 11
felle (n pl) **87**, 12
fellhersteller (m pl) **83**, 16
fertigbreite (f) **83**, 4
fester Industriekord (m) **55**, 5
feststellen **10**, 13
fett/fettig (adj) **84**, 13
feucht (adj) **88**, 4
feuchtigkeit (f) **121**, 13
feuchtigkeitsgehalt (m) **121**, 14
feuchtigkeitsmesser (m) **121**, 15
feuerbeständiges Tuch (n) **35**, 8
feuerfestes Gewebe (n) **80**, 16
feuerfester Stoff (m) **35**, 10
feuergefährlich, brennbar (adj) **46**, 9
feuergefährlich, brennbar (adj) **88**, 14
fibres **152**, 6
fiiermaschine (f) **57**, 14
fiiermaschine (f) **57**, 14
filato (m) – lana inglese **157**, 11
filialleiter (m) **13**, 17
filterstoff (m) **35**, 6
filzbildung (f) **81**, 16
finish **77**, 1
firma (f) **50**, 2
fläche (f), Oberfläche (f) **9**, 13
flächengewicht (n) **9**, 14
flachs (m) **83**, 7
flachs (m)/Leinen (n) **83**, 8
flachstrickerei (f) **83**, 6
flammenbeständiges futter (n) **91**, 12
flammenfeste Faser (f) **82**, 3
flammenfestes Gewebe (n) **81**, 1
flammenfestes Mittel (n) **83**, 5
flaschengrün (adj) **19**, 12
fleck (m) **137**, 14
flocke- und Lumpenhändler (m pl) **83**, 10
flockefärben (n) **138**, 11
flockehersteller (m pl) **83**, 9
flordecke (f) **127**, 9
flordeckenhersteller (m pl), Plüsch-Hersteller (m pl) **127**, 11
florstoff (m) **38**, 13
fluchtlinie (f), Ausrichten (n) **6**, 4
fluoreszierende Aufheller (n pl) **83**, 12
flüssig (adj) **92**, 2
for woven and knitted piece manufacturers **50**, 18
flüssigkeit (f) **83**, 11
flüssigkeit (f) **92**, 1
förderband (n) **54**, 8
förderbandanlage – Maschinen (f pl) **101**, 3
förderbandeinrichtung (f) **54**, 9
förderbandgitter (n) – Maschinen (f pl) **101**, 4
förderbandsystem (n) **54**, 10
forderungspreis (m) **10**, 14
forschung (f) – Textilforschung (f) **131**, 11
frachtbrief (m) **16**, 3
frachtfrei (adj) **26**, 2
freizeitbekleidung **73**, 10
freizeitbekleidung **78**, 1
freizeitbekleidungstoff (m) **81**, 3
frist (f), Verspätung (f), Verzögerung (f) **62**, 17
frottieren, reiben **60**, 1
frobewolle (f), -fäden (m pl), -stoffe (m pl) **79**, 13
fubekleidungsstoffe (m pl) **35**, 15
führen (v) **85**, 4
füllfaser (f) **82**, 1
füllfaser für – Spielzeug (m) und weiche Heimtextilien (f pl) **82**, 2
füllung (f)/Schu (m) **149**, 2
futter (n) – Karo/Karofutter (n) **91**, 11
futter (n) – Polyesterfutter (n) **91**, 15
futter (n) – jacquard, Jacquardfutter (n) **91**, 13
futterstoff (m) **37**, 13

G

gabardinestoff (m) **35**, 17
garantie (f) **85**, 3
garn (n) **153**, 10
garn (n) – Abfallgarn (n)/Garnabfall (m) **163**, 12
garn (n) – Acrylbaumwollgarn (n) **153**, 15
garn (n) – Acrylendlosgarn (n) **154**, 4
garn (n) – Acrylflammeneffektgarn (m) **154**, 5
garn (n) – Acrylgarn (n) **153**, 11
garn (n) – Acrylgarnmischungen (f pl) **153**, 12

karomuster (n), Würfelmuster (n) **28**, 4
karren und Schienenwagen (m pl) **143**, 17
kartenbindemaschine (f) **25**, 6
kartenstanze (f) **25**, 7
karton (m), Gehäuse (m), Schachtel (f) **26**, 4
karton (m) **24**, 12
kaschiermaschinen (f pl) **91**, 3
kaschiertes Tuch (n) **31**, 17
kaschmir (m) **26**, 8
kaschmir (m) – enthaart **85**, 16
kaschmir (m) – gereinigt **85**, 18
kaschmirhaar (n) **85**, 12
kaschmirkämmlinge (m pl) **123**, 8
kaschmirmantelstoffe (m pl) **41**, 6
kaschmirstoff (m) **32**, 17
kaseinfaser (f) **26**, 6
kastanienbraun (adj) **11**, 6
kasten (m) **26**, 5
katalog (m) **26**, 10
katalysator (m) **26**, 11
katalysatoren (m pl) **26**, 12
kattunstoff (m) **32**, 12
kaufbrief (m) **16**, 4
kaufen (n) **22**, 8
käufer (m) **22**, 6
käufermarkt (m) **22**, 7
kavallerietwill-Wollstoff (m) **33**, 4
kavallerietwillstoff (m) gemischt **33**, 3
kegel (m), Konus (m); Spule (f), konische Kreuzspule (f), Kone (f) **51**, 12
kerngarn (n), Coregarn (n) **55**, 7
kettbaum (m) **15**, 4
kette (f) **27**, 7
ketten (f pl) – Schlichtenmittel (f pl) **145**, 18
ketten (f pl) – Wirker (m pl) **146**, 1
kettenantrieb (m) **27**, 8
kettengleitmittel (n pl) **146**, 2
kettenknüpferei (f) **146**, 7
kettenwachs (n) **146**, 3
kettenwickel für die Mopindustrie **13**, 9
kettenwirkerei (f) **146**, 6
kettfäden (m pl) in gleichen Abständen **79**, 9
kettschlichter (m pl) **145**, 17
khaki (adj) **90**, 2
kirschfarben (adj) **27**, 5
kklagen **50**, 6
klären, Läuterung (f), Reinigung (f) **31**, 4
klärung (f) **30**, 14
klassierung (f), Sortierung (f) **30**, 15
klassierung (f), Sortierung (f) **30**, 17
klebstoffe (m pl) **4**, 7
kleid (n), Anzug (m) **67**, 13
kleid (n) **67**, 12
kleiderstoff (m) **67**, 18
kleiderstoff (m) **80**, 13
kleidung (f) (adj) **8**, 13
kleidung (f), Bekleidung (f) **36**, 9
klette (f) **22**, 2
klettwolle (f) **22**, 3
klimatisierung (f) **5**, 11
knick (m) **59**, 9
knitterfest (adj) **8**, 4
knitterfest (adj) **57**, 17
knitterfest (adj) **58**, 1
knittererholung (f) **57**, 18
knoten (m), Knötchen (n) **122**, 12
knoten (m) **90**, 11
knotenlos (adj) **90**, 3
knüpfen (m) **90**, 13
koagulierungsmittel (n) **43**, 1
kobaltblau (adj) **43**, 12
koeffizient (m) **43**, 18
kohäsion (f) **44**, 1
kohlendioxyd (n) **23**, 16
kohlenhydrat (n) **23**, 15
kohlenoxyd (n), Kohlenmonoxide (n pl) **24**, 8
kohlensäure (f) **23**, 17
kokon (m) **43**, 17
kollodal (adj) **44**, 9
kolloid (n) **44**, 8
kolorimeter, Farbmesser (m) **44**, 10
kolorimetrisch (adj) **44**, 11
kommission (f), Auftrag (m) **46**, 12
kommission (f) Lohnsortierer (m pl) **48**, 12
kommission (f) für Garnbürster (m pl) Lohn-Gernbürster (m pl) **49**, 7
kommission (f) für Garnnumerierer (m pl), Lohn – Garnschlichter (m pl) **49**, 14
kommission (f) für Kämmer (m pl) Lohnkämmer (m pl) **47**, 5
kommission (f) für Mischer (m pl) **46**, 14
kompensation (f) **50**, 4
komponente (f) **50**, 8
kondensat (n) **51**, 5
kondensation (f), Kondensierung (f) **51**, 6
kondensieren, verdichten **51**, 7
konditionieren (n), Klimatisieren (n), Konditionierung (f) **51**, 10
konerei (f) **52**, 4
konferenz (f) **51**, 15
konisch, kegelförmig **52**, 3
konitionierkammer (f) **51**, 11
konkav (adj) **51**, 2
konkurrenz (f) **50**, 5
konsistenz (f) **51**, 9
konsortium (n) **52**, 11
konten (n pl) **2**, 5
konisch, kegelförmig **51**, 3
kontinentales Spinnen (n) **53**, 2
kontinue-Waschmaschine (f) **53**, 11
kontinuefärbemaschine (f) **53**, 7
kontoauszug (m) **14**, 2
kontolle (f) **54**, 2
kontinuierlich, endlos (adj) **53**, 3
kontinuierliche Spinnmachine (f) **52**, 12

L

M

N

O

P

S

s-drehung (f) **139**, 5
sack (m) **12**, 16
sackleinwan (f) **12**, 17
sacknähzwirne (m pl) **133**, 11
sakrale Bekleidung (f), Priesterbekleidung (f) **33**, 8
salpetersäure (f) **123**, 1
salz (n) **133**, 15
sammlung (f), Kollektion (f) **44**, 7
samstoff (m) **42**, 9
sargtuch (n) **33**, 10
satin (m) **133**, 18
sauber, rein **31**, 1
sauerfett (n) **2**, 18
saugen, ansaugen **67**, 8
saugfähig **1**, 14
säurebad (n), Säureflotte (f) **3**, 4
säurebeständigkeit (f), Säurefestigkeit (f) **3**, 5
säureechtheit (f) **2**, 17
säurefarbstoff (m) **2**, 16
säurelösung (f) **3**, 6
säuren (f pl) **3**, 3
schädlich, gefahrvoll, gefährlich (adj) **60**, 17
schaftstuhl (m) **66**, 3
schals (m pl) **134**, 1
schalsstoff (m) **40**, 1
schalwollstoff (m) **39**, 16
schären (n) **146**, 5
schärmaschine (f) **146**, 4
schattierung (f) **134**, 11
schattierungskarte (f), Farbkarte (f) **134**, 12
scheck (m) **29**, 4
scheren (n) **60**, 7
scheren **134**, 15
scheuerfestigkeit (f) **1**, 5
scheuerprüfgerät (n) **1**, 7
schieben (n), Schlupf (m), Kriechen (n) **58**, 6
schiffchen (n), Schützen (m) **135**, 5
schimmel (m) **121**, 3
schlange (f), Windung (f) **44**, 2
schläuche (m pl) – Kunststoff (m) **144**, 2
schläuche (m pl) – flach **144**, 1
schlauchförmig **99**, 9
schlechter Auenstand (m) **12**, 14
schleierstoff (m) **42**, 7
schleifmittel (n pl) **1**, 9
schleuder (f), Zentrifuge (f) **88**, 7
schlichter (m pl) **135**, 9
schlingenbildung (f) **136**, 10
schmälzen (f pl) für Faser **82**, 4
schmelzen (n) **120**, 13
schmiermittel (n pl) – Fette (n pl) und Öle (n pl) **92**, 7
schmiermittel (n pl) – Leim-Schlichte **92**, 8
schmiermittel (n pl) – Mischung (f) **92**, 5
schmiermittel (n pl) – Trockenpulver (n) **92**, 6
schmiermittel (n pl) für Garn **92**, 10
schmiermittel (n pl) für Textilien **92**, 9
schmutzgarn **84**, 16
schmutzig (adj) **65**, 1
schmutzstoff (m) **84**, 14
schmutzwolle (f) **84**, 15
schneidemaschine (f) **60**, 9
schnitt (m), Schneiden (n), Scherung (f) **60**, 1
schnitt (m), Schneiden (n), Scherung (f) **60**, 6
schnittlänge (f) **60**, 2
schnüre (f pl) und Fäden (m pl) **55**, 4
schotenkaro (n) reiner-Wollstoff (m) **40**, 14
schottenkar-Mischungstoff (m) **40**, 13
schrumpfen (n) **135**, 3
schrumpffest **135**, 2
schrumpfschutzbehandlung (f) **135**, 4
schufaden (m) **127**, 5
schulden (f pl) **61**, 14
schuldner (m) **61**, 15
schurwolle (f) **145**, 9
schürzentuch (n) **31**, 15
schüsse pro cm **127**, 6
schustreifigkeit **14**, 10
schütteln **5**, 2
schütteln (n), Umrühen (n) **5**, 3
schüttelvorrichtung **117**, 8
schutzschicht (f) **130**, 3
schützenloser Webstuhl **135**, 6
schuzähler (m) **127**, 8
schwarz (adj) **16**, 6
schwarzes Schaf (n) **16**, 8
schwarzköpfiges Schaf (n) **16**, 7
schwer(adj) **64**, 10
seefrachtspediteur (m) **83**, 15
seide (f) **135**, 7
seidenkämmlinge (m pl) **123**, 16
seidenstoff (m) **40**, 4
seidenzwirner (m)/Seidenspinner (m) **141**, 2
seife (f) **136**, 11
seifen (f pl) – textil **136**, 13
seifen (f pl) **136**, 12
seil (n), Strang (m), Schnur (f), Bindfaden (m) **54**, 18
seil (n), Strang (m) **133**, 3
seitlich, Seiten- **44**, 6
semikammgarnspinnen (n)/Halbkammgaon-spinnen (m) **134**, 8
shoddy (n); wertlos **134**, 17
shoddyhersteller (m pl) **134**, 18
sicherheitsprodukt (n) **133**, 12
sieden **19**, 9
sieden (m) **19**, 10
sitzungszimmer (n) **19**, 6
sortieren (n) **137**, 4
spannrahmentrocknung (f) **140**, 13

wollkammgarnspinner (m pl) - Spezialfasern (f pl) **153**, 2
wollkammgarnspinner (m pl) - Super 80 plus **153**, 1
wollkämmlinge (m pl) **123**, 17
wollkarbonisierer (m pl) **24**, 6
wollkarden (f pl) - Maschinen (f pl) **97**, 4
wollkaschmirstoff (m) **33**, 2
wollmittel (n pl) **149**, 12
wollseidestoff (m) **40**, 5
wollsergen (f pl) **39**, 18
wollsiegelrockstoff (m) **40**, 8
wolltuchhersteller (m pl) - Mischwolle/Wollsiegel **152**, 1
wolltuchhersteller (m pl) - Spezialfasern **152**, 2
wolltuchhersteller (m pl) - Wollsiegel **152**, 3
wolltuchhersteller (m pl) **151**, 16
wundnähfäden (m pl) **139**, 8

Y

yak (n) **153**, 9
yakhaar (n) **86**, 14
yakkämmlinge (m pl) **123**, 18

Z

z-drehung (f) **165**, 9
zarte Farben (f pl), flüchtige Farben (f pl), Signierfarben (f pl) **141**, 3
zellophan (n) **26**, 14
zellulose (f), Cellulose (f), Zellstoff (m) **26**, 15
zellulosefaser (f) **26**, 16
zentigrad (m) **62**, 8
zentigramm (n) **26**, 18
zentimeter (m) **27**, 1
zentrieren (n) **26**, 17
zentrifugal (adj) **27**, 2
zentrifugalextraktor (m) **27**, 3
zentrifuge (f) **27**, 4
zerfallen, zersetzen **62**, 5
zerlegan, abbauen **62**, 6
zerstäuber (m) **11**, 3
ziegenhaar (n) **86**, 5
ziegenhaarkämmlinge (m pl) **123**, 11
ziegelrot (adj) **20**, 6
zieheranfälligkeit (f), Fadenziehen (m) **136**, 9
zimtbraun (adj) **30**, 2
zoll (m) **59**, 16
zoll (m) **69**, 3
zollerklärung (f) **59**, 17
zollfrei (adj) **69**, 5
zolltarif (m) **59**, 18
zubehörteile (n pl) **2**, 3
zugfestigkeit (f) **140**, 12
zulässig (adj) **6**, 12
zulieferer an Hausmeister **89**, 9
zureichen (n), Eintichen (n) **131**, 12
zusammensetzung (f), Komposition (f) **50**, 9
zusammenpressen **50**, 11
zussammenziehen, falten **130**, 8
zusatz (m), Zusetzen (n) **4**, 2
zusatz (m) **4**, 3
zuschneiden (n) **60**, 8
zusetzen, nachsetzen **4**, 1
zustand (m), Bedingung (f), Kondition (f) **51**, 8
zweifach-, Doppel-, **144**, 9
zwirnmachine (f) **67**, 1
zwirnstoff (m) Zwirntuch (n) **42**, 3
zylinder (m) **60**, 11

Espanol
Spanish
Spanisch
Espagnol
Spagnolo

agentes (m pl) de protección contra polillas **122**, 1
agentes (m pl) de telas **31**, 13
agentes (m pl) fluorescentes para resaltar **83**, 12
agentes (m pl) igualadores de color **69**, 9
agentes (m pl) para acondicionar hilo **155**, 14
agentes (m pl) para reforzar hilo **163**, 1
agents (m pl) de seguros **89**, 1
agitación (f) **5**, 3
agitador (m) **5**, 4
agitar **5**, 2
aglomeración (f) **2**, 6
agua (f) **148**, 9
agua (f) caliente **88**, 3
agua (f) destilada **65**, 14
agua (f) fría **44**, 5
aguja (f) **122**, 9
agujas (f pl) **122**, 11
aire (m) **5**, 6
aire (m) caliente **88**, 1
aire (m) comprimido **50**, 12
aislador (m) **88**, 17
ajustar, adaptar **4**, 9
alambre (m) de cardas, alambre (m) de guarnición de carda **25**, 12
alambre (m) de lizo **68**, 2
albaricoque (m) **9**, 5
alcalino **6**, 8
alcanzar (m) **131**, 12
alcohol (m) **6**, 2
aleación **6**, 14
alfombra **25**, 13
alfombrillas (f pl) de cachemir, lana y mezclas **133**, 9
algodón (m) **55**, 14
algodón (m) cardado **24**, 18
algodón (m) de coser **61**, 3
algodón (m) hidrófilo **1**, 13
algodón (m) hidrófilo **57**, 3
algodón (m) peinado **46**, 1
algodón/poliéster **73**, 9
alineación (f) **6**, 4
alizarina (f) **6**, 5
almacén (m) **145**, 14
almacenaje (m) **138**, 16
almacenaje (m) de toda clase de fibra **138**, 17
almacenaje (m) **145**, 15
alteración (f) **6**, 16
alternante **6**, 17
amarillo **165**, 7
amarillo canario **23**, 9
ambarino (adj) **6**, 18
amianto (m) **10**, 9
amilopectina (f) **7**, 7
aminoácido (m) **7**, 3
amoladores (m pl) de encargo, amoladores por encargo **47**, 13
amoníaco (m) **7**, 4
amonio (m) **7**, 5
añadir **4**, 1
análisis (m) **7**, 8
análisis (m) de los tejidos, examen (m) de las telas **40**, 18
análisis (m) químico **28**, 13
analistas (m pl) **7**, 9
analítico **7**, 10
analizar **7**, 11
anchura (f) **149**, 7
anchura (f) abierta **125**, 5
anchura (f) ancha **20**, 11
anchura (f) de peine **131**, 16
anchura (f) final **83**, 4
anchura (f) total **125**, 11
angola **7**, 13
ángulo (m) **7**, 12
anhídrido (m) carbónico, dióxido (m) de carbono **23**, 16
anilina (f) **7**, 15
anillos (m pl) de cera **148**, 15
anillos (m pl) de maquinaria de hilar **114**, 7
animal (m) **7**, 16
anticongelante (m) **8**, 6
antideslizante **8**, 8
antiestático **8**, 10
antiinflamable **91**, 12
anudar (m) **90**, 13
aparato (m) **8**, 12
aparatos (m pl) de interrumpir el funcionamiento de maquinaria de hacer punto **115**, 2
apelotonadores (m pl) **13**, 8
aplastamiento (m) **59**, 9
aplicación (f) **8**, 17
aplicar a a **8**, 18
aprendiz (m), aprendiza (f) **9**, 2
aprensadores (m pl), calandrado (m) **23**, 5
aprestadores (m pl) **135**, 9
aprestar (m) **135**, 8
aprestar, encolar **67**, 14
apresto (m), acabado (m) **67**, 16
apropiado **9**, 3
aproximado **9**, 4
área (f) **9**, 13
arpillera (f) **12**, 17
arrancadores (m pl) de residuos **146**, 17
arrancar (m) la lanzadera **13**, 11
arreos (m pl) **87**, 4
arrolladores (m pl) de goma **133**, 1
arruga (f) **153**, 7
arruga (f), doblez (m), pliegue (m), pliegue (m) de planchado **57**, 16
arrugar (m) **43**, 16
arsénico (m) **10**, 1
artículo (m) **10**, 2

cáñamo (m) **87**, 10
cáñamo (m), yute (m) **89**, 13
canillera (f) **54**, 16
caño (m) de salida **89**, 11
cantidad (f) **130**, 12
capa (f) **128**, 3
capacidad (f), potencia (f), energía (f) **23**, 11
capas (f pl) de protección **130**, 3
capas (f pl) impermeables de tela de pura lana virgen **41**, 10
capas (f pl) impermeables para tela **41**, 4
capas (f pl) impermeables para tela de cachemir **41**, 6
capas (f pl) impermeables para tela de fibras naturales **41**, 8
capas (f pl) impermeables para tela de mezclas de lana/cachemir etc. **41**, 9
capas (f pl) impermeables para tela de pelo de camello **41**, 5
capas (f pl) impermeables para tela mezclada **41**, 7
capas (f pl) para forros de pantallas oscuras **43**, 7
capas (f pl) para persianas enrollables **43**, 8
capilar **23**, 12
capullo (m) **43**, 17
caqui (m) **90**, 2
característica (f), propiedad (f) **27**, 13
carbonato (m) de calcio **22**, 14
carbonización (f) **23**, 18
carbonización (f) **27**, 17
carbonizadores (m pl) de lana **24**, 6
carbonizadores (m pl) de residuos **24**, 5
carbonizadores (m pl) de telas **24**, 3
carbonizadores (m pl) de trapos **24**, 4
carbonizar **24**, 1
carbonizar (m) **27**, 12
carbonizar (m), carbonización **24**, 7
carda (f) **24**, 10
cardado **24**, 15
cardadores (m pl) **24**, 16
cardadores (m pl) por encargo **46**, 17
cardadores (m pl) por encargo **47**, 5
cardadura (f) **134**, 4
cardar **24**, 11
cardar (m) algodón **56**, 2
cardas (f pl) de alambre **140**, 7
cardencha (f) **140**, 4
cardenchas (f pl) **140**, 6
cardón (m), cardado (m), cardadura (f) **25**, 3
carga (f) **27**, 14
cargar, alimentar, aprestar **27**, 15
caro **61**, 11
carrete (m) **127**, 13
carrete (m) **137**, 13
carrete (m) mecha **141**, 4
carretillas (f pl) para pelos **105**, 10
carro (m) **26**, 1
carros (m pl) para tintorerías **69**, 8
carros (m pl) para transportar lotes grandes **97**, 6
carros (m pl) y carretillas (f pl) **143**, 17
carta (f) de matices **134**, 12
cartón (m) **24**, 12
casa (f) de acondicionar de Bradford **19**, 18
casa (f) pública de acondicionamiento **130**, 6
catalizador (m) **26**, 11
catalizadores (m pl) **26**, 12
catálogo (m) **26**, 10
caudal (m) de aire **5**, 14
celofán **26**, 14
celulosa (f) **26**, 15
centígramo (m) **26**, 18
centímetro (m) **27**, 1
centraje (m) **26**, 17
centrífugo **27**, 2
centrífugo (m) **27**, 4
cepilladores (m pl) de hilos por encargo **49**, 7
cepillar (m) **21**, 6
cepillo (m) **20**, 14
cepillos (m pl) **20**, 15
cepillos (m pl) de fibra carta desechables **21**, 1
cepillos (m pl) de toda clase, de alambres, torcidos tubos – neumáticos etc. **20**, 16
cepillos (m pl) disponibles de fibras cortas **21**, 2
cepillos (m pl) industriales **21**, 4
cepillos (m pl) para la industria de alfombras **20**, 17
cepillos (m pl) para teñir y acabar **21**, 3
cepillos para tejidos – fabricados y reparados **21**, 5
cera (f) **148**, 13
cera (f) animal **8**, 2
ceras (f pl) para urdimbre **146**, 3
cerda (f) artificial **10**, 4
certificado **132**, 2
certificados (m pl) de calidad **27**, 6
chamuscar **134**, 2
chartreuse (m) **27**, 18
chenilla (f), felpilla (f) **29**, 3
cheque (m) **29**, 4
chiffón (m) **29**, 7
chinchillá (f) **29**, 9
chintz (m), zaraza (f), glaseda (f) **29**, 10
cilindro (m) **60**, 11
cinta (f) **132**, 12
cinta (f) de carda **25**, 10
cinta (f) transportadora **54**, 8
cintas (f pl) **139**, 13
cintas (f pl) adhesivos **139**, 14
cintas (f pl) autoadhesivas **139**, 15
cintas (f pl) de polipropileno **139**, 16

D

E

G

H

hilo (m) de estambre para hacer punto **164**, 17
hilo (m) de estambre para tejer **165**, 3
hilo (m) de estambre teñido **164**, 15
hilo (m) de estambre y lana teñidos y mezclados **165**, 1
hilo (m) de fantasía **158**, 2
hilo (m) de felpilla **155**, 12
hilo (m) de fibras exóticas/lana **161**, 6
hilo (m) de filamentos **157**, 13
hilo (m) de franjas **163**, 2
hilo (m) de ganchillo **156**, 16
hilo (m) de hacer punto **158**, 7
hilo (m) de hacer punto a mano **158**, 7
hilo (m) de hacer punto de aran **154**, 12
hilo (m) de importación **158**, 12
hilo (m) de lana **163**, 17
hilo (m) de lana **164**, 4
hilo (m) de lana blanca **164**, 11
hilo (m) de lana británica **155**, 1
hilo (m) de lana de cordero **159**, 2
hilo (m) de lana de cordero/angora/nilón **159**, 3
hilo (m) de lana inglesa **157**, 10
hilo (m) de lana mezclada **164**, 5
hilo (m) de lana natural **164**, 9
hilo (m) de lana para calcetería **164**, 7
hilo (m) de lana para hacer punto **164**, 8
hilo (m) de lana para tejer **164**, 10
hilo (m) de lana pura para alfombras **155**, 6
hilo (m) de lana pura y clasificada **164**, 12
hilo (m) de lana regenerada **162**, 8
hilo (m) de lana teñida **164**, 6
hilo (m) de lana/algodón **163**, 18
hilo (m) de lana/nilón acrílico **164**, 2
hilo (m) de lana/nilón **164**, 1
hilo (m) de lana/viscosa **164**, 3
hilo (m) de lienzo **159**, 4
hilo (m) de lienzo/algodón **159**, 5
hilo (m) de lienzo/lana **159**, 10
hilo (m) de lienzo/viscosa **159**, 8
hilo (m) de lino hilado con agua – hilera y estopa **159**, 9
hilo (m) de lino hilado en seco – hilera y estopa **159**, 6
hilo (m) de lino para hacer punto a máquina **159**, 7
hilo (m) de lurex **159**, 12
hilo (m) de mechones **162**, 13
hilo (m) de mechones acrílicos **154**, 5
hilo (m) de mechones de algodón **156**, 9
hilo (m) de merino/extra **160**, 4
hilo (m) de mezclas acrílicas **153**, 12
hilo (m) de mezclas de algodón **155**, 18
hilo (m) de mezclas de lana para alfombras **155**, 7
hilo (m) de mezclas de poliéster **160**, 18
hilo (m) de monofilamentos **160**, 9
hilo (m) de nilón **160**, 12
hilo (m) de nilón/lana **160**, 14
hilo (m) de piel **157**, 17
hilo (m) de pelo **158**, 5
hilo (m) de pelo de cabra **158**, 4
hilo (m) de pelo de camello **155**, 3
hilo (m) de pelo de camello/lana **155**, 4
hilo (m) de poliéster con algodón de extremo abierto **156**, 7
hilo (m) de poliéster **160**, 17
hilo (m) de poliéster para coser **162**, 3
hilo (m) de poliéster/lana **161**, 2
hilo (m) de poliéster/viscosa **161**, 1
hilo (m) de polipropileno **161**, 3
hilo (m) de rayón hilado **162**, 15
hilo (m) de residuos **163**, 12
hilo (m) de rizo **159**, 11
hilo (m) de seda **162**, 9
hilo (m) de seda hilada **162**, 16
hilo (m) de seda/lana **162**, 11
hilo (m) de seis capas **162**, 12
hilo (m) de tejer de cheviot **155**, 13
hilo (m) de torcer **163**, 8
hilo (m) de trama **127**, 5
hilo (m) de trenzas mercerizadas **160**, 1
hilo (m) de vidrio **158**, 3
hilo (m) de viscosa **163**, 10
hilo (m) de yak **165**, 6
hilo (m) dintético/lana **163**, 6
hilo (m) donegal **157**, 3
hilo (m) dralón **157**, 5
hilo (m) elástico **157**, 9
hilo (m) elástico **162**, 18
hilo (m) especial de algodón **156**, 10
hilo (m) fino de visón **157**, 14
hilo (m) grueso **43**, 4
hilo (m) hilado con aguja hueca **165**, 5
hilo (m) hilado en torno de hilo nuclear **155**, 15
hilo (m) hiladores **162**, 14
hilo (m) ignífugo **157**, 16
hilo (m) industrial **158**, 13
hilo (m) ligero de Shetland **162**, 7
hilo (m) mate acrílico y regular **154**, 2
hilo (m) merino **160**, 3
hilo (m) mestizo para calcetería **156**, 17
hilo (m) mezclado – medio estambre **161**, 15
hilo (m) mezclado **154**, 17
hilo (m) mulato para tejer **157**, 1
hilo (m) natural **160**, 11
hilo (m) para alfombras **155**, 5
hilo (m) para alfombrillas **161**, 10
hilo (m) para aljofifas **160**, 10
hilo (m) para calcetería **158**, 11
hilo (m) para cintas **161**, 9
hilo (m) para cortinas **157**, 2
hilo (m) para gabardinas **158**, 1
hilo (m) para la exportación **157**, 11
hilo (m) para mantas **154**, 16

I

J

L

M

N

O

P

Q

R

S

tela (f) no tejida **38**, 9
tela (f) no teñida **42**, 4
tela (f) para abrigos de caballeros **38**, 1
tela (f) para abrigos de senoras **37**, 5
tela (f) para almacenes **42**, 12
tela (f) para arrugarse **33**, 9
tela (f) para banderas **35**, 9
tela (f) para bufandas **39**, 15
tela (f) para calzados **35**, 15
tela (f) para chales **40**, 1
tela (f) para chaquetas **36**, 12
tela (f) para chaquetas ligeras **32**, 7
tela (f) para colchas **80**, 7
tela (f) para corbatas **38**, 8
tela (f) para cotrinas **80**, 12
tela (f) para decoración **81**, 2
tela (f) para estampación **81**, 9
tela (f) para estolas **39**, 13
tela (f) para fajas **33**, 15
tela (f) para faldas **40**, 6
tela (f) para forrar corbatas **41**, 3
tela (f) para forros tiesos e internos **36**, 11
tela (f) para forros tiesos **36**, 5
tela (f) para gabardinas **35**, 17
tela (f) para gorras y sombreros **32**, 16
tela (f) para jipijapas **38**, 12
tela (f) para juguetes **41**, 12
tela (f) para manguitos **9**, 7
tela (f) para mantas de viaje **41**, 13
tela (f) para mantas **32**, 3
tela (f) para máquinas de lavanderías **37**, 7
tela (f) para mesas de billar y trucos **32**, 2
tela (f) para muebles **35**, 16
tela (f) para pantalones **42**, 1
tela (f) para paracaídas **81**, 6
tela (f) para pelotas de tenis **40**, 15
tela (f) para persianas **42**, 11
tela (f) para ropa informal **37**, 8
tela (f) para ropa interiores y femeninas **81**, 4
tela (f) para ropas informal **81**, 3
tela (f) para saxony [tejido fino] **39**, 14
tela (f) para sillas plegables **34**, 9
tela (f) para tapicería **81**, 13
tela (f) para tapicerías **42**, 6
tela (f) para toallas **41**, 11
tela (f) para trajes de estambre **40**, 12
tela (f) para trajes de lana **40**, 11
tela (f) para trajes de lienzo **40**, 10
tela (f) para trajes deportivos **40**, 9
tela (f) para trajes **33**, 16
tela (f) para uniformes **42**, 5
tela (f) para velas de yates **81**, 12
tela (f) para velos **42**, 7
tela (f pl) para forrar féretros **33**, 10
tela (f) plegable **35**, 14
tela (f) prensada con vapor **32**, 8
tela (f) prensada **38**, 18
tela (f) recortada **34**, 6
tela (f) rehumedecida **34**, 8
tela (f) rematada con agua **42**, 15
tela (f) rematada en ambos lados **34**, 14
tela (f) rematada en seco **34**, 16
tela (f) remendada **37**, 17
tela (f) resistente a las lanas **81**, 1
tela (f) resistente al fuego **35**, 8
tela (f) resistente al fuego **80**, 16
tela (f) reversible **39**, 7
tela (f) shetland **40**, 3
tela (f) tejida **42**, 14
tela (f) teñida **69**, 11
tela (f) tipo pana bedford **32**, 1
telar (m) ancho **20**, 10
telar (m) con pinzas **85**, 2
telar (m) con toberas de aire **5**, 17
telar (m) de levas **140**, 2
telar (m) de toberas de agua **148**, 10
telar (m) de vapor **89**, 12
telar (m) dobby **66**, 3
telar (m) jacquard **89**, 7
telar (m) manual **86**, 17
telar (m) para alfombras **25**, 16
telar (m) sin lanzadera **135**, 6
telares (m pl) manuales **105**, 11
telares (m pl) para telas pesadas e industriales **108**, 2
telas - convertidores **38**, 4
telas (f pl) bordadas **80**, 15
telas (f pl) coordenadas **80**, 10
telas (f pl) de contrato **33**, 13
telas (f pl) de estambre fino **89**, 10
telas (f pl) estrechas **122**, 6
telas (f pl) forradas **31**, 17
telas (f pl) jacquard **89**, 6
telas (f pl) para cubiertas de estampación **81**, 8
telas (f pl) para delantales **31**, 15
telas (f pl) para los cuellos tipo melton **33**, 11
telas (f pl) para maquinaria de papel y cartón **81**, 5
telas (f pl) para vestidos sacerdotales **33**, 8
telas (f pl) resistentes a arrugarse – algodón, viscosa y mezclas **80**, 11
temperatura (f) **140**, 9
teñido a trechos **137**, 5
teñido bicolor **59**, 5
teñido en bobina **125**, 14
teñido en corona **22**, 12
teñido (m) de fibras sueltas no elaboradas **138**, 11
teñido (m) de piezas **127**, 7
teñido (m) en mecha **141**, 5
teñidura (f) de madejas **87**, 2
teñir, colorar **44**, 14
teñir dos veces **59**, 4
teñir (m) a presión **129**, 7
teñir (m) con cromo **29**, 18

tintoreros (m pl) y rematadores (m pl) de algodón/poliéster **73**, 7
tintoreros (m pl) y rematadores (m pl) de algodón/viscosa **73**, 11
tintoreros (m pl) y rematadores (m pl) de angora **75**, 10
tintoreros (m pl) y rematadores (m pl) de encogidos **76**, 11
tintoreros (m pl) y rematadores (m pl) de géneros de algodón **73**, 5
tintoreros (m pl) y rematadores (m pl) de géneros de punto **77**, 3
tintoreros (m pl) y rematadores (m pl) de géneros de rizo **77**, 2
tintoreros (m pl) y rematadores (m pl) de géneros de vellón **74**, 4
tintoreros (m pl) y rematadores (m pl) de géneros para decoración **74**, 5
tintoreros (m pl) y rematadores (m pl) de hilo de algodón **74**, 1
tintoreros (m pl) y rematadores (m pl) de impermeabilizantes **76**, 10
tintoreros (m pl) y rematadores (m pl) de impermeabilizantes **77**, 9
tintoreros (m pl) y rematadores (m pl) de lana/Poliéster **77**, 12
tintoreros (m pl) y rematadores (m pl) de lana/algodón unidos **77**, 10
tintoreros (m pl) y rematadores (m pl) de lana/estambre **77**, 11
tintoreros (m pl) y rematadores (m pl) de lienzo **74**, 10
tintoreros (m pl) y rematadores (m pl) de lienzo /viscosa **75**, 2
tintoreros (m pl) y rematadores (m pl) de lienzo/algodón unidos **74**, 11
tintoreros (m pl) y rematadores (m pl) de lienzo/poliéster **75**, 1
tintoreros (m pl) y rematadores (m pl) de mantas de algodón **73**, 3
tintoreros (m pl) y rematadores (m pl) de mezclado **72**, 14
tintoreros (m pl) y rematadores (m pl) de mezclado y sintético **72**, 15
tintoreros (m pl) y rematadores (m pl) de mezclas de poliéster/algodón **76,3**
tintoreros (m pl) y rematadores (m pl) de mezclas de poliéster/viscosa **76**, 4
tintoreros (m pl) y rematadores (m pl) de mezclas de viscosa/algodón **77**, 5
tintoreros (m pl) y rematadores (m pl) de poliéster **76**, 1
tintoreros (m pl) y rematadores (m pl) de poliéster puro (100%) **76**, 2
tintoreros (m pl) y rematadores (m pl) de prendas hechas de punto **74**, 9
tintoreros (m pl) y rematadores (m pl) de ropa de trabajo y ropa informal **78**,1
tintoreros (m pl) y rematadores (m pl) de ropa de trabajo y ropa informal hechas de algódon/poliéster **73**, 9
tintoreros (m pl) y rematadores (m pl) de ropa impermeable de algódon/poliéster **73**, 8
tintoreros (m pl) y rematadores (m pl) de viscosa **77**, 4
tintoreros (m pl) y rematadores (m pl) de viscosa/lino **77**, 6
tintoreros (m pl) y rematadores (m pl) de viscosa/poliéster **77**, 7
tintoreros (m pl) y rematadores (m pl) de viscosa/poliéster y lino **77**, 8
tintoreros (m pl) y rematadores (m pl) **72**, 12
tintoreros (m pl) y rematadores (m pl) para forros tiesos e internos **74**, 8
tintoreros (m pl) y rematadores (m pl) resistentes a la putrefacción **76**, 8
tintoreros (m pl) y rematadores (m pl) termofijación **74**, 6
tintoreros y rematadores maquinaria para lotes de muestras **75**, 4
tintoreros y rematadores maquinaria para tejidos lavables de lana **75**, 7
tintoreros y rematadores maquinaria para telas tejidas **75**, 8
tintoreros y rematadores maquinaria para viscosa y telas maquinaria para viscosa resistente al arrugado **75**, 6
tintura (m) de hilo **157**, 7
tiradores (m pl) de cáñamo **89**, 15
tiradores (m pl) de encargo – 4 cilindros La Roche **48**, 4
tiradores (m pl) de encargo **48**, 3
tiradores (m pl) de residuos de lana pura **147**, 11
tiradores (m pl) de residuos de lana/sintéticos **147**, 14
tiradores (m pl) **130**, 9
tiritas (f pl) **136**, 1
tiza (f)/yeso (m) **27**, 9
tizas (f pl) fugitivas para tejidos **27**, 10
toldo (m) **12**, 5
tolerancia (f) **6**, 13
tolva (f) **87**, 14
torcedores (m pl) de hilo por encargo **49**, 15
torcedores (m pl) de seda **141**, 2
torcedura (f) S **139**, 5
torcedura (f) Z **165**, 9
torcer (m) lana **136**, 5
torsión (f) **144**, 8
torsión (f) al revés **132**, 10
torsión (f) suave **136**, 18
traje (m), vestido (m) **67**, 12
traje (m), vestido (m) **67**, 13
traje (m), vestido (m), ropa (f) **67**, 10
trama (f) **149**, 2

Francais
French
Französisch
Francés
Francese

B

contrôle (m) **54**, 2
contrôle (m) **88**, 15
convenable **9**, 3
conversion (f), transformation (f) **54**, 6
convertir **54**, 7
coordonnée **54**, 13
corde (f), boyau (m), cordon (m), ficelle (f) **54**, 18
corderie (f) **133**, 5
cordes (f pl) et fils (m pl) retors **55**, 4
cordes (f pl) pour l'industrie lourde **55**, 5
corrélation (f) **55**, 8
correspondre **5**, 5
correspondre **55**, 9
costume (m), habit (m) **67**, 10
côte (f) **55**, 1
coton (m) à coudre **61**, 3
coton (m) à polir **146**, 13
coton (m) cardé **24**, 18
coton (m) écru **33**, 18
coton (m) **55**, 14
coton (m) hydrophile, ouate (f) hydrophile **1**, 13
coton (m) peigné **46**, 1
cotonnade (f) **32**, 12
cotonnerie (f) **56**, 1
coton/polyester **73**, 9
couches (f pl) de protection **130**, 3
couleur (f) **44**, 13
coupe (f) au hasard – commission **48**, 5
coupeuse (f), machine (f) à couper **60**, 9
coupon (m) d'échantillon **139**, 9
courant (m) d'air **5**, 12
courbé, courbe **59**, 2
courroie (f) de commande **68**, 1
courroie (f) **15**, 14
courroie (f) trapézoïdale **145**, 4
courroies (f pl) **15**, 16
courroies (f pl) pour convoyeurs **53**, 9
courtelle **57**, 12
courtiers (m pl) d'assurance **89**, 1
courtiers (m pl) en laine **149**, 18
coussinet (m) **15**, 7
coût (m) **55**, 10
coutil (m) pour literie **41**, 2
couverture (f), housse (f) **16**, 9
couvertures (f pl) de voyage **143**, 15
craie (f) **27**, 9
craies (f pl) fugaces [textiles] **27**, 10
crayons (m pl) – fugitifs/fugaces **57**, 15
créance (f) **58**, 2
créancier (m) **58**, 3
crêpe (m) de Chine **58**, 8
crêpe (m) **58**, 7
cretonne (f) **58**, 10
crin (m) artificiel **10**, 4
crispage (m) de tissu **33**, 9
critique **58**, 18
croisé **59**, 6
croisé lourd – laine **33**, 4
croisé lourd – mélangé **33**, 3
cru **131**, 5
cuir (m) **91**, 4
cuve (f) **145**, 2
cylindre (m) **60**, 11
cylindres (m pl) en caoutchouc **133**, 1

D

dacron (m) **60**, 12
damas (m) **60**, 14
date (f) limite **61**, 6
de couleur d'améthyste **7**, 1
de laine **78**, 7
de peu de poids **91**, 10
débit (m) **61**, 12
débiteur (m) **61**, 15
débouillisage (m) en continu **53**, 10
débourreurs (m pl) de laine **150**, 5
décarboniser **61**, 16
décatissage (m) London shrink **92**, 3
décatissage (m) [de tissu] **34**, 8
d'échantillon **75**, 4
déchets (m pl) – alpaga **146**, 9
déchets (m pl) – arracheuses – laine/synthétiques **147**, 14
déchets (m pl) – arracheuses – nappes et filature **147**, 13
déchets (m pl) – découpeuses **146**, 14
déchets (m pl) – délaineurs purls laines **147**, 11
déchets (m pl) – fil de tapis **146**, 10
déchets (m pl) – pinces arracheuses pour bonneterie **147**, 12
déchets (m pl) – polyester **147**, 7
déchets (m pl) – pour récupération synthétique **146**, 16
déchets (m pl) de bonneterie **146**, 18
déchets (m pl) de carde **25**, 11
déchets (m pl) de coton **146**, 12
déchets (m pl) de fil **163**, 12
déchets (m pl) de filature et bonneterie **148**, 1
déchets (m pl) de filé **148**, 8
déchets (m pl) de laine **148**, 6
déchets (m pl) de laine peignée **148**, 7
déchets (m pl) de mohair **147**, 3
déchets (m pl) de nylon **147**, 4
déchets (m pl) de polypropylène **147**, 8
déchets (m pl) de soie **147**, 15
déchets (m pl) de tissu **42**, 13
déchets (m pl) d'extraction **80**, 4
déchets (m pl) **146**, 8
déchets (m pl) **60**, 10
déchets (m pl) plastiques **147**, 6
déchets (m pl) synthétiques **148**, 2
déchirer (v) **140**, 3

E

équipement (m) en matériel de finissage (m) des tricots **107**, 6
équipement (m) en matériel de foulage du drap **98**, 8
équipement (m) en matériel de foulage du drap **99**, 1
équipement (m) en matériel de foulage **108**, 10
équipement (m) en matériel de foulardage **110**, 7
équipement (m) en matériel de fusion **105**, 2
équipement (m) en matériel de lainage **115**, 7
équipement (m) en matériel de lissage **98**, 1
équipement (m) en matériel de mercerisage **108**, 8
équipement (m) en matériel de mesure d'humidité **108**, 13
équipement (m) en matériel de mise en balles du tissu **97**, 8
équipement (m) en matériel de montage des garnitures **95**, 10
équipement (m) en matériel de rattachement de fil **119**, 10
équipement (m) en matériel de récupération des fibres **104**, 7
équipement (m) en matériel de teinture de tissu **98**, 5
équipement (m) en matériel de thermofixage de fil **119**, 9
équipement (m) en matériel de thermofixage **106**, 2
équipement (m) en matériel de vérification des tissus **98**, 6
équipement (m) en matériel de vérification du tissu **103**, 7
équipement (m) en matériel d'emballage **110**, 4
équipement (m) en matériel d'encollage de la cha9ne **117**, 1
équipement (m) en matériel d'enroulage du tissu **97**, 9
équipement (m) en matériel d'ensimage **109**, 10
équipement (m) en matériel d'ensimage (m) des fibres **104**, 6
équipement (m) en matériel d'ensouplage **94**, 3
équipement (m) en matériel d'étirage **102**, 7
équipement (m) en matériel d'extrusion **103**, 6
équipement (m) en matériel d'impression hectographique **116**, 4
équipement (m) en matériel d'occasion **113**, 6
équipement (m) en matériel fait sur demande **101**, 10
équipement (m) en matériel hydraulique **106**, 7
équipement (m) en matériel métiers (m pl) continus **112**, 3
équipement (m) en matériel modernisé **108**, 12
équipement (m) en matériel non-automatique **109**, 6
équipement (m) en matériel peralta **110**, 9
équipement (m) en matériel pneumatique **111**, 5
équipement (m) en matériel pour carton de commande **95**, 11
équipement (m) en matériel pour la teinturerie **103**, 1
équipement (m) en matériel pour le floc (m) et le chiffon (m) **104**, 10
équipement (m) en matériel pour le laboratoire de teinturerie **103**, 2
équipement (m) en matériel pour le non-tissé (m) **109**, 8
équipement (m) en matériel pour les métiers et le tissage **107**, 11
équipement (m) en matériel pour moquettes et tapis **97**, 5
équipement (m) en matériel pour ouvrir les balles **93**, 3
équipement (m) mesure de la largeur **119**, 2
essai **11**, 5
essai (m) **140**, 14
essai (m) **143**, 16
et de loisir **78**, 1
établissement (m) du prix de revient **55**, 12
étaler, présenter **65**, 11
étamine (f) **22**, 5
etc **94**, 1
étiquette (f) **90**, 14
étiquettes (f pl) **90**, 15
étoffe (f) de coton **33**, 17
étoffe (f) de coton non-teinte **34**, 2
étoffe (f) de crin **36**, 3
étoffe (f) de laine de confection **40**, 11
étoffe (f) de laine peignée de confection **40**, 12
étoffe (f) de laine peignée/mohair/trevire **41**, 14
étoffe (f) de laine peignée/trevire **41**, 15
étoffe (f) de laine/cachemire/mélanges etc. pour paletot **41**, 9
étoffe (f) de laine/coton **42**, 18
étoffe (f) d'habillement **80**, 13
étoffe (f) industrielle **36**, 8
étoffe (f) lin **37**, 9
étoffe (f) matelassée **34**, 13
étoffe (f) nappée **38**, 9
étoffe (f) pour casquettes et chapeaux **32**, 16
étoffe (f) pour machines de blanchisserie **37**, 7
étoffe (f) pour tablier **31**, 15
étoffe (f) pour tablier **9**, 7

F

fil (m) **153**, 10
fil (m) – de laine peignée **164**, 13
fil (m) Aran à tricoter **154**, 12
fil (m) Berber **154**, 15
fil (m) Saxony **161**, 13
fil (m) Shetland **162**, 4
fil (m) à âme **55**, 7
fil (m) à coudre **134**, 9
fil (m) à coudre **162**, 2
fil (m) à coudre polyester **162**, 3
fil (m) à crochet **156**, 16
fil (m) à poil **160**, 15
fil (m) à tisser croisé **157**, 1
fil (m) à tisser **164**, 14
fil (m) à tricot cueilli **164**, 15
fil (m) à tricoter à la main Aran **158**, 8
fil (m) à tricoter à la main **158**, 7
fil (m) à tricoter coloré **158**, 18
fil (m) à tricoter **158**, 17
fil (m) à tricoter très doux **164**, 4
fil (m) abrasé texturé à la filature (f) **1**, 3
fil (m) acier inoxydable **162**, 17
fil (m) artificiel **159**, 13
fil (m) avec boutons noueux **159**, 1
fil (m) bouclé **159**, 11
fil (m) brodeur **158**, 2
fil (m) câblé **22**, 10
fil (m) cardé **25**, 2
fil (m) cardé super **164**, 3
fil (m) cashmilon **155**, 10
fil (m) cha9nette **155**, 11
fil (m) chenille **155**, 12
fil (m) cheviote **155**, 13
fil (m) continu **157**, 13
fil (m) continu **53**, 15
fil (m) courtelle **156**, 14
fil (m) d'acrylique boutonneux **154**, 5
fil (m) d'acrylique brillant gonflé **153**, 13
fil (m) d'acrylique brillant régulier **153**, 14
fil (m) d'acrylique **153**, 11
fil (m) d'acrylique gonflé **154**, 1
fil (m) d'acrylique mat régulier **154**, 2
fil (m) d'acrylique mélangé **153**, 12
fil (m) d'acrylique nylon **154**, 3
fil (m) d'acrylique open end **154**, 4
fil (m) d'alpaga **154**, 8
fil (m) d'alpaga/laine **154**, 9
fil (m) d'amiante **154**, 13
fil (m) d'angola **154**, 10
fil (m) d'angora **154**, 11
fil (m) d'artisanat **156**, 15
fil (m) d'artisanat pour macramé **158**, 6
fil (m) de Jersey **158**, 15
fil (m) de cachemire **155**, 8
fil (m) de coton acrylique **153**, 15
fil (m) de coton boutonneux **156**, 9
fil (m) de coton cardé **156**, 1
fil (m) de coton **155**, 16
fil (m) de coton **56**, 16
fil (m) de coton **57**, 4
fil (m) de coton imitation **156**, 3
fil (m) de coton mélangé **155**, 18
fil (m) de coton mercerisé **159**, 16
fil (m) de coton open end **156**, 6
fil (m) de coton open end/polyester **156**, 7
fil (m) de coton peigné **156**, 2
fil (m) de coton shetland noueux **162**, 5
fil (m) de coton spécial **156**, 10
fil (m) de coton teint **156**, 4
fil (m) de coton/acrylique **155**, 17
fil (m) de coton/laine **156**, 12
fil (m) de coton/nylon **156**, 5
fil (m) de coton/polyester **156**, 8
fil (m) de coton/viscose **156**, 11
fil (m) de déchets de soie **162**, 16
fil (m) de fouet **42**, 16
fil (m) de garniture, fil (m) pour garniture de carde **25**, 12
fil (m) de laine à six brins pour tapis **161**, 12
fil (m) de laine acrylique **154**, 6
fil (m) de laine britannique **155**, 1
fil (m) de laine cardée à tisser **164**, 10
fil (m) de laine cardée à tricoter **164**, 8
fil (m) de laine cardée blanc **164**, 11
fil (m) de laine cardée coloré **164**, 6
fil (m) de laine cardée **164**, 4
fil (m) de laine cardée mélangé **164**, 5
fil (m) de laine cardée naturel **164**, 9
fil (m) de laine cardée pour bonneterie **164**, 7
fil (m) de laine cardée woolmark **164**, 12
fil (m) de laine d'agneau **159**, 2
fil (m) de laine d'agneau/angora/nylon **159**, 3
fil (m) de laine peignée à tisser **165**, 3
fil (m) de laine peignée à tricoter **164**, 17
fil (m) de laine peignée blanc **165**, 4
fil (m) de laine peignée coloré **164**, 15
fil (m) de laine peignée mélangé **164**, 14
fil (m) de laine peignée mi-blanc **165**, 2
fil (m) de laine peignée mi-coloré **165**, 1
fil (m) de laine peignée mi-mélangé **164**, 18
fil (m) de laine peignée pour bonneterie **164**, 16
fil (m) de laine simple pour tapis **161**, 11
fil (m) de laine/cachemire **155**, 9
fil (m) de laine/coton **164**, 18
fil (m) de laine/fibres rares **161**, 6
fil (m) de laine/nylon **164**, 1
fil (m) de laine/nylon/acrylique **164**, 2
fil (m) de laine/poil de chameau **155**, 4
fil (m) de laine/soie **162**, 11
fil (m) de laine/synthétique **164**, 6
fil (m) de laine/viscose **164**, 3
fil (m) de lana inglesa **157**, 10
fil (m) de lin – tricot machine **159**, 7
fil (m) de lin **159**, 4

G

H

I

M

Q

R

S

T

U

V

Y

Italiano
Italian
Italiensch
Italiano
Italien

amminoacido (m) **7**, 3
ammissibile **6**, 12
ammoniaca (f) **7**, 4
ammonio (m) **7**, 5
ammorbidente (m) **136**, 15
ammorbidenti (m pl) **136**, 16
ammorbidenti (m pl) per stoffe **81**, 11
amoladores (m pl) de encargo, amoladores por encargo **47**, 14
analisi (f) **7**, 8
analisi (f) chimica **28**, 13
analisti (m pl) **7**, 9
analitico **7**, 10
analizzare **7**, 11
anelli (m pl) di cera **148**, 15
angolo (m) **7**, 12
angolo (m) di inserzione chiuso, passo (m) chiuso **31**, 7
angora (f) **7**, 14
anidride (f) carbonica, biossido (m) di carbonio **23**, 16
anilina (f) **7**, 15
animale (m) **7**, 16
annodare (v) l'ordito (annoclare dell'ordito) **146**, 7
annuario (m), guida **64**, 19
anti putrefazione/resistente alla putrefazione **133**, 6
antigelo (m) **8**, 6
antipiega **8**, 4
antipiega **58**, 1
antisdrucciolevole **8**, 8
antistatico **8**, 10
antitarme **121**, 18
apparecchi (m) **8**, 12
apparecchiature (f) da laboratorio ed esame/prove **91**, 1
apparecchiature (f) per la salute e la sicurezza **87**, 6
apparecchiature (f pl) di depurazione ad aria **5**, 9
applicare a **8**, 18
applicare un colore su un altro **125**, 12
applicazione (f) **8**, 17
apprendista (m) **9**, 2
apprettare **67**, 14
apprettatrici (f pl) dell'ordito/apprettatori dell'ordito **145**, 17
appretto (m), finissaggio (m) **67**, 16
apprezzabile notevole, significatiuo **9**, 1
appropriato, adatto **9**, 3
approsiusiuo – tivo **9**, 4
apriballe (m) **13**, 5
apritoi di lana **151**, 1
apritoi (m pl) – per cascame **125**, 2
apritoi (m pl) – per lana **125**, 3
apritoi (m pl) – per sintetici **125**, 1
apritoio (m) **124**, 17
arazzo (m) **9**, 18
archeggio (m), flessione (f) **19**, 16
area (f) zona (f), superficie (f) **9**, 13
aria (f) **5**, 6
aria (f) calda **88**, 1
aria (f) compressa **50**, 12
armadietti (m pl) per l'abbinamento di colori **45**, 13
armatura tela **127**, 15
armatura diagonale **64**, 6
armatura (f) a nido d'ape **87**, 13
armatura (f) diagonale **144**, 6
armatura (f) **148**, 17
armatura (f) panama **14**, 12
arresto (m) automatico [telaio] **138**, 15
arricciare **59**, 13
arricciatura (f) **43**, 16
arricciatura (f), crettatura (f), increspatura (f) **58**, 15
arsenico (m) **10**, 1
articolo (m) **10**, 2
artificiale **10**, 3
asciugamento (m) **68**, 8
aspetto (m) **8**, 14
aspirare, assorbire **67**, 8
assegno (m) **29**, 4
assemblea (f) del consiglio di amministrazione **19**, 5
assenteismo (m) **1**, 10
assenza (f), mancanza (f) **1**, 8
assorbente **1**, 14
assorbente (m) **1**, 12
assorbimento (m) **1**, 15
assorbire **1**, 11
asta (f) **11**, 7
astrakan (m) **10**, 17
atmosfera (f) **10**, 18
attaccare **11**, 4
attivare **3**, 12
attivatore (m) **3**, 13
attività (f) **3**, 14
atto (m) di cessione, atto di vendita **16**, 4
attrezzatura (f) tessile elettronica **79**, 6
ausiliari (m pl) del colorante **69**, 16
ausiliari (m pl) di pelle **91**, 5
ausiliario (m) **10**, 16
ausiliario (agg) **11**, 18
autocarri (m pl) e carrelli (m pl) **143**, 17
autoclave (f) **11**, 10
automatico **11**, 13
automazione (f) **11**, 17
avvolgimento (m) del tessuto sotto tensione, crabbing (m) [fissaggio a pumido] **57**, 13
avvolgimento (m) su coni **52**, 4
azocolorante (m) **12**, 8
azzurro **12**, 9

B

bagnato **149**, 4

D

E

F

J

K

L

M

Q

R

S

scarto (m) – per rigenerazione sintetica **146**, 16
scarto (m) – pettinatura **146**, 11
scarto (m) – pulitura del cotone **146**, 13
scarto (m) – strofinacci **148**, 4
scarto (m) – tagliatrici **146**, 14
scarto (m) – tiratoi **147**, 10
scarto (m) – tiratoi – lana/sintetici **147**, 14
scarto (m) – tiratoi – pinzette da maglieria **147**, 12
scarto (m) – tiratoi – tela di battitoio e filatura **147**, 13
scarto (m) – tiratoi per lana pura **147**, 11
scarto (m) dei commercianti **147**, 2
scarto (m) di calzetteria (f) **146**, 18
scarto (m) di filato **148**, 8
scarto (m) di filatura e maglieria **148**, 1
scarto (m) di garnettatrici **146**, 17
scarto (m) di lana **148**, 5
scarto (m) di materie sintetiche **148**, 2
scarto (m) di pettinato **148**, 7
scarto (m) di plastica **147**, 6
scarto (m) di poliestere **147**, 7
scarto (m) di polipropilene **147**, 8
scarto (m) di selezionatrici **147**, 16
scarto (m) di seta **147**, 15
scarto (m) di trasformatrici **147**, 9
scarto (m) d'importatori **147**, 1
scarto/cascame **146**, 8
scatola (f) **26**, 4
scatole (f pl) di cartone, scatole **24**, 13
sciarpe (f pl) **134**, 1
scienza (f) degli ultrasuoni/ultrasuoni **144**, 11
scolorare, scolorire **65**, 3
scoloritura (f) del colore **45**, 6
scomporre **62**, 6
sconto (m) **65**, 6
scorrimento (m) **58**, 6
scorrimento (m)/slittamento **135**, 15
scorta (f), riserva (f), accumulo (m) **12**, 11
scuro **61**, 1
sdoganamento (m) d'importazione **88**, 10
sdoganamento (m) **31**, 3
secchezza (f) **68**, 10
secco **68**, 4
secco ad aria **5**, 13
segno (m) del taglio, segno (m) della pezza **60**, 3
segno (m) di dente del pettine/rigature del pettine (f pl) **131**, 15
selezionatrici di lana **151**, 8
selezione (f) **134**, 5
semi-automatico **134**, 7
senso (m) **64**, 17
serbatoio (m), deposito (m) **52**, 17
serie (f) di carde, assortimento di carde **25**, 9
serpentino (m), spira (f) **44**, 2
servizi (m pl) di computer ed elaborazione di testi **51**, 1
servizi (m pl) di traduzione **143**, 12
servizi (m pl) d'interpretazione **89**, 2
servizi video **145**, 8
servizio (m) **69**, 4
servizio (m) divisione **63**, 10
seta (f) **135**, 7
seta (f) artificiale **10**, 8
seta (f) cruda **59**, 8
seta (f) grezza **131**, 7
setola (f) artificiale **10**, 4
sfilacciatura (f) degli orli [tessitura] **81**, 15
sfilacciatura (f) di stracci **130**, 14
sfregare **59**, 1
sfumatura (f) **134**, 13
sgrassare **62**, 15
sintetico **119**, 14
sintetico **139**, 10
sistema (m) cotoniero **56**, 15
sistema (m) decimale **62**, 3
sistema (m) di controllo **54**, 4
sistemi (m pl) di controllo **54**, 5
sistemi (m pl) di controllo automatizzato **11**, 12
sistemi (m pl) di controllo computerizzati per produttori di tessuto e maglieria
sistemi (m pl) di conversione del vapore **138**, 8
sistemi (m pl) di movimentazione dei materiali **120**, 8
sistemi (m pl) di riferimento di colori **45**, 16
sistemi (m pl) di ventilazione **145**, 6
sistemi (m pl) trasportatori **54**, 10
smagliatura (f) **136**, 9
società (f) **50**, 2
soda (f) caustica **26**, 13
soffiamento **19**, 1
solfato (m) di calcio **23**, 1
solfato (m) di sodio **136**, 14
solidità (f) agli acidi **2**, 17
solidità (f) agli alcali **6**, 7
solidità (f) del colore, solidità (f) tintoriale **45**, 7
soluzione (f) **137**, 1
soluzione (f) acida **3**, 6
soluzione (f) alcalina **6**, 9
solvente (m) **137**, 2
solventi (m pl) **137**, 3
sopporto (m), cuscinetto (m) **15**, 7
sostanze (f pl) chimiche per il settore tessile **29**, 1
sostituto (m) **139**, 6
sovratingere **59**, 4
sovratintura (f) **59**, 5
spaghi (m pl) **144**, 7
spazzola (f) **20**, 14
spazzolatura **21**, 6
spazzole (f pl) **20**, 15

T

U

V

Y

ELLIOT MUSGRAVE LTD.
Jackson Street, Bradford
West Yorkshire, BD3 9SJ U.K.
Telephone: 0274 731115
Telefax: 0274 722691

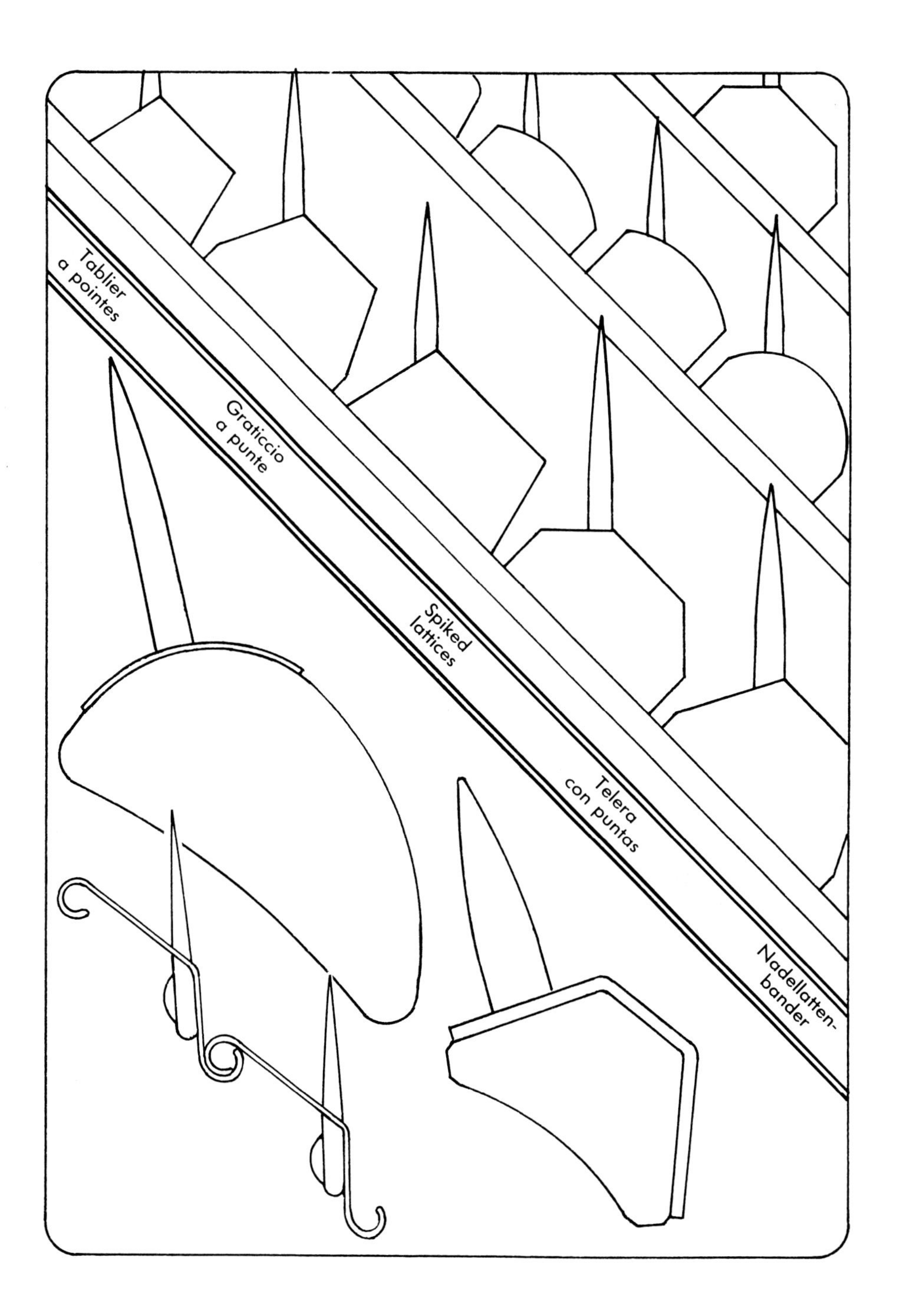
Tablier a pointes
Graticcio a punte
Spiked lattices
Telera con puntas
Nadellatten-bander

Samuel Lumb & Son Ltd.

Perseverance Mills, Elland, West Yorkshire HX5 9AX.
Tel: (0422) 373434 (5 lines). Telex: 51414 SLS G. Fax: (0422) 375747

Woollen spinners of speciality yarns for tweeds — upholstery — wall coverings — hand knitting, in 100 % natural fibres. Count range 0,5 nm to 10 nm.

Bayer UK Ltd.

4 James Street, Bradford, West Yorkshire BD1 3PZ.
Tel: (0274) 390736. Telex: 51205. Fax: (0274) 305904

Producers of Acrylic fibre, tow and tops for furnishings, knitwear and hand knitting. Elastomeric yarns for foundation and swimwear.
Nylon fibre for floor coverings and industrial fabrics.
Nylon monofils and acrylic multifilament for industrial fabrics.
TRADE NAMES: Dralon, Dorlaston, Dorix, Perlon, Dunova.

CRABTREE

David Crabtree & Son Ltd.

Dick Lane, Bradford, West Yorkshire BD4 8JD.
Tel: (0274) 662251. Telex: 51412. Fax: (0274) 656074

Gripper-Jacquard and Gripper-Spool Axminster carpet weaving machinery; Ancillary Machines; Spare parts for Craig Axminster and Wilton looms.

Courtaulds Fibres Ltd

Courtelle, Westcroft Mill, Great Horton, Bradford, West Yorkshire.
BD7 3EL
Tel: (0274) 571 151. Telex: 51494. Fax: (0274) 521208

Suppliers of 100 % Courtelle blended tops to the hand knitting and machine knitting domestic market and also export markets.
100 % Coloured Neochrome Courtelle tops.
100 % Courtelle Ecru tops.
Blends of Courtelle with Natural and speciality fibres.
Blends of Courtelle with synthetic and effect fibres.
Also commission topmakers.

TRADE NAMES: COURTELLE, NEOCHROME.

Cobble Blackburn Limited

Gate Street, Blackburn BB1 3AH.
Tel: (0254) 55121. Telex: 63159. Fax: (0254) 671125.

Textile machinery manufacturers including: carpet tufting machines, pattern attachments, carpet backing plants.
BRANCHES: Wilson & Longbottom; Pep/Pendhill.

Herdmans Ltd.,

Sion Mills, Co. Tyrone. BT82 9HE.
Tel: (06626) 58421. Telex: 74626 HLY SM G. Fax: (06626) 58909.

Wetspun linen yarns from: 14 LEA to 140 LEA, 8.5 NM to 84 NM.
TRADE NAME: Herdmans Ltd.
BRANCHES: Herdmans (Ireland) Ltd., Ballybofey, Co. Donegal, Ireland.

Sandoz Chemicals (A Division of Sandoz Products Ltd.)
Calverley Lane, Horsforth, Leeds, LS 18 4RP
Tel: (0532) 584646. Telex: 557114 SDZ LDS. Fax: (0532) 390063

One of the world's leading manufacturers and suppliers of dyestuffs and chemical auxiliaries for use in preparation, Dyeing, Printing and Finishing of all Textiles. **Dyes for Cellulosic Fibres** — Drimarene X, X-N, K, R & P, Indosol, Solar, Diresul RTD. **For Nylon** — Nylosan "E", "N" & "F" Lanasyn. **For Polyester** — Foron "E", "RD", "S" & "SE". **For Wool** — Drimalan, Lanasan, Lanasyn, Omega-Chrome and Sandolan "E" MF and Milling. **Chemical Brands** — Preparations: Sandozin, Stabilizer, etc. Dyeing Assistants: Dilatin, Fadex, Lyogen, Sandacid, etc. Finishing: Cerannine, Leucophor (Optical Brightening), Sandlube, Sandofluor, etc.
BRANCHES: Mandervell Rd, Oadby Industrial Estate, Leicester.

Breaks Bros. Ltd.
Powell Road, Bolton Woods, Shipley BD18 1BD,
Tel: (0274) 583241, Telex: 666729,

Textile auxiliary manufacturers of products for the scouring and dyeing, for all classes of fibres particularly for woollen and worsted manufacturers.

TRADE NAMES: Chlorethol, Exenol, Emenol, Medenol, Sofsil, Sepron, Targol, Unex, Vanol, Zenol.

BRANCH: Newbridge Chemicals Works, Radcliffe, Manchester.

Jerome Fabrics

S. Jerome & Sons Ltd., Victoria Works, Shipley, West Yorkshire BD17 7EF.
Tel: (0274) 587251. Telex: 51685. Fax: (0274) 588654.

Manufacturers of designed Suiting, Trouserings and Jacketings wool — wool blends and synthetics for home and export.

Yarn Aid Engineers Ltd.
Mortar Street, Oldham, Lancashire. OL4 2BA
Tel: 061 624 5600. Telex: 665139 YARAD G. Fax: 061 626 1853.

Manufacturers of high precision rotor units and wired opening rollers for all current and most obsolete open-end spinning machines. Full maintenance/refurbishing workshop including rewiring with dynamic balancing of rotating parts. Workshop facilities include 3 balancing machines (Schenck & Reutlinger), Hardinge Super Precision Toolroom Lathe and Tsugami CNC 3 axis turning centres. Inspection equipment includes profile projector and air and electronic gauging by Mercer & Mitutoyo.

P. W. Greenhalgh Co. Ltd.,
Ogden Mill, Newhey, Nr. Rochdale, Lancs. OL16 3TH.
Tel: (0706) 847911, Telex: 635067, Fax: (0706) 881217.

Independent Commission Bleachers, Dyers Finishers, Fusible coaters and Back-coaters of Cotton and Pe/C Fabrics. Specialising in the acrylic foam back thermal coating for the curtain and curtain lining trade, and fusible coatings on interlinings for the shirt, blouse and clothing trade.

TRADE NAME: SUNKING.

F. Harding (Macclesfield) ltd.
Kershaw Mill, Newton Street, Macclesfield, Cheshire. SK11 6QZ.
Tel: (0625) 29625. Telex: 666131. Fax: (0625) 612836.

Throwsters — all types of yarn processing, spun, continuous filament, monofilament, winding and doubling — heat setting etc. Dyed Yarns supplied.

The Elton Cop Dyeing Co. Ltd.
Walshaw Road, Bury, Lancashire BL8 1NQ,
Tel: 061-764 1383, Telex: 669810, Fax: 061-763 1159,

Commission dyers and merchants of mercerised cotton, soft cotton, polyester, polyamides, blends and fancy yarns. All yarns dyed on package in the companys' automated dyehouse.